Why Science?

Why Science?

JAMES TREFIL

Teachers College, Columbia University
New York and London

National Science Teachers Associatrion
Arlington, VA

Published simultaneously by Teachers College Press, 1234 Amsterdam Avenue, New York, NY 10027 and by the National Science Teachers Association (NSTA), 1840 Wilson Boulevard, Arlington, VA 22201.

Library of Congress Cataloging-in-Publication Data

Trefil, James S., 1938–
  Why science? / James Trefil.
    p. cm.
Includes bibliographical references and index.
ISBN 978-0-8077-4830-5 (pbk.) —ISBN 978-0-8077-4831-2 (hardcover)
1. Science—Study and teaching. I. Title.
  Q181.T8195 2007
  507.1—dc22
2007018130

ISBN: 978-0-8077- 4830-5 (paper)
ISBN: 978-0-8077- 4831-2 (hardcover)

Printed on acid-free paper
Manufactured in the United States of America

15 14 13 12 11 10 09 08   8 7 6 5 4 3 2 1

In a book devoted to education, it is fitting that I dedicate this work to my two daughters, each of whom has finished the formal part of hers:

Dominique Waples-Trefil, J.D., University of Denver

Flora Waples-Trefil, M.D., Cornell University

Contents

Preface

SCIENTIFIC LITERACY

This is a book about science education and, more explicitly, about the aspect of science education that concerns itself with people who are not going to be scientists. Historically, most educators' attention has been focused on finding ways to improve the national supply of technically competent men and women—an important goal. Today, however, faced with national issues that are increasingly acquiring scientific and technical dimensions, we are starting to turn to the question of how to go about providing average citizens with enough scientific knowledge to allow them to participate in public debates in a meaningful way. In the language of this book, this means that we are trying to produce citizens who are scientifically literate.

This is a complex subject. I begin in Chapter 1 with a basic discussion of the nature of science—what it is and, perhaps more important, what it is not. Once the reader has an understanding of the way this particular field of intellectual effort works, I turn to the first task—a definition of *scientific literacy*. In Chapter 2 this question is approached in the broader context of what has come to be called *cultural literacy:* The body of knowledge that educated people, at a given time and in a given place, assume other people possess. In this scheme, *scientific literacy* is seen to be the knowledge about science that the average citizen needs in order to take part in the public debates that are so important in a democracy. To take a standard example, someone who wishes to participate in the current debate about the use of embryonic stem cells needs to know enough biology to understand what a stem cell is, what an embryo is, and what the pros and cons of using embryonic stem cells are. Similar kinds of background knowledge can be defined for any public debate with a scientific or technological component. For this book, then, I define *scientific literacy* as the matrix of knowledge needed to understand enough about the physical universe to deal with issues that come across the horizon of the average citizen, in the news or elsewhere.

Having defined scientific literacy, I then use the next three chapters to present arguments as to why it is important. In Chapter 3 I expand the argument given above, that citizens actually need this sort of knowledge to function in democracy. I title this the "Argument from Civics." In Chapter 4 I point out that science is an important part of culture, a part that average citizens ought to know about for the same reason that they ought to know something about history or literature. I title this the "Argument from Culture." This subject leads to the standard Two Cultures debate; I talk about recent new wrinkles in this debate coming from postmodern philosophy. Finally, in Chapter 5 I take up the point that understanding nature can add to our enjoyment of the world we live in. I title this the "Argument from Aesthetics." Among other subjects, I look at the standard "we murder to dissect" statements about the incompatibility of science and art (or, perhaps more accurately, between science and Art).

Having established the desirability of producing a scientifically literate citizenry, I then turn to the question of the state of scientific literacy in America and the rest of the world. In Chapter 6 I present the history of data collection on this issue and discuss the difficult question of how one goes about measuring a quantity like scientific literacy. In essence, social scientists look at two things: whether people understand the constructs of science (what is the difference between an atom and a molecule?) and the methods of science (what is a control group?).

One of the problems one always encounters when scientists talk about scientific literacy is a tendency to blur the difference between science and technology. In Chapter 7, therefore, I examine this relationship in some detail, presenting the viewpoint of scientists involved in applied research and development.

The next two chapters are devoted to an examination of the history of science education and an analysis of our current state of affairs—where we are and how we got here. The bottom line is that it is only quite recently that people have started to be concerned about the state of scientific knowledge in the population at large. The consequences of the publication of *A Nation at Risk*, the rise of Physics for Poets classes at universities, and the current scientific literacy movement are all traced out in Chapter 8. In Chapter 9 I examine the current institutional barriers to improving scientific literacy in our educational system, from middle school to universities.

Having discussed where we are, I then turn to the question of where we ought to go. In Chapter 10 I look at the question of what the goals of science education for nonprofessionals ought to be. This is a subject on which reasonable people can—and do—differ, so I outline the various points of view first and then argue that scientific literacy constitutes a minimal goal in any educational scheme. In Chapter 11 I support this argument by talking about the way the science is changing because of the availability of the computer. The science of tomorrow (and even, for that matter, some of the science of today) doesn't look like the traditional kind of discipline around which our educational system has been built—another argument for change in that system.

Finally, in Chapter 12, I present a scheme for organizing the teaching of scientific literacy. This scheme is built around the internal structure of science itself. Intellectually, science can be thought of as a kind of hierarchy, with important results and concepts flowing naturally from a relatively small number of what I like to call Great Ideas. These grand, overarching principles form a kind of skeleton for the scientific worldview. They unify nature and tie it together. Consequently, they also form an ideal basis for a general education in science.

I argue that students who have acquired a working knowledge of these Great Ideas will be equipped to deal with the scientific aspects of public issues. They will, in short, be scientifically literate.

FOR WHOM IS THIS BOOK INTENDED?

The short answer to this question is simple: This book is intended for anyone interested in science education. More particularly, it is intended for people involved in planning the science education experience for that great (but often neglected) majority of students who will not pursue careers in science and technology. These students do not need to be turned into miniature scientists, but they do need to know enough science to be able to function as citizens in later life.

I invite educators at all levels, from kindergarten to graduate school—teachers, administrators, and curriculum designers—to join me in a discussion of this crucial topic.

ACKNOWLEDGMENTS

Many people have helped me by sharing their insights and reading all or part of the manuscript of this book. With the proviso that any errors that remain are my responsibility alone, I would like to thank Robert Blonski, Janina Blonski, Elizabeth Falk, Ryszard Michalski, Jon Miller, Jeff Newmeyer, Richard Okreglak, Valdimir Petrushevsky, Martin Piecuch, and Jerzy Sapeyevski. Most of all, I would like to thank my wife, Wanda, for many useful discussions and for introducing me to the real world of K–12 education. I would also like to thank Brian Ellerbeck of Teachers College Press for suggesting the topic of this book and shepherding it through production, and Wendy Schwartz for exceedingly useful editorial suggestions. Finally, I would like to thank the staff of Kiari's Coffee House in Fairfax, Virginia, for providing the pleasant atmosphere and gallons of good cappuccino it took to finish the writing.

CHAPTER 1

Science: A World Understood

Why science?
Why do it? Why think about it? Why teach it or study it? And why, for God's sake, write a book about it?

In a sense, the rest of this book will be an extended answer to these questions, but I will start by reminding you of how recent the human entry into a science-dominated world has been. For most of the time our species has been on the planet, we lived in what Carl Sagan called the "Demon-Haunted World"—a world in which mysterious, unpredictable, and often malevolent forces dominated human understanding. And while that world has not entirely receded from sight (think of the way many people regard cancer), the natural world today seems more manageable, more predictable, somehow safer than it did to our ancestors. When you understand that lightning is the result of the movement of charged particles rather than the wrath of the gods, it changes the way you think about it. A world understood is a world less threatening, and understanding the world is what, in the end, science is about.

But you don't have to get philosophical to see the effects of science and technology in your life. Take yourself through a typical day. Did you wake up to a radio alarm? Radio waves were predicted in 1861, discovered in 1888, and used to send a signal in 1894. Did you drive to work? Then you started your car with a battery. The first battery was built in 1800. The laws of thermodynamics that govern the operation of your car's engine were first laid out in 1842, and the first modern internal combustion engine ran in 1876. Did you turn on a computer on your desk when you got to work? The basic laws that govern the materials in that computer were discovered in the 1930s, the first primitive computer built in the 1940s.

I could go on, but I think the point is clear. The material things that define our lives, as well as our rational-technological view of the world, are all relatively recent results of the scientific enterprise. Because we live in the scientific world, it's easy for us to overlook the true magnitude of what science has done in the short time it's been around.

Look at it another way. Imagine that you were an extraterrestrial approaching Earth for the first time. You would immediately notice one thing. Of the millions of species of life, only one—*Homo sapiens*—has developed the ability to alter the environment to suit its needs; only one has learned enough about the laws that govern the universe to affect the functioning of an entire planet. Natural landscapes everywhere have been altered to grow crops useful to humans while huge cities have sprung up to house them. No other species comes close to having the impact that we do. And whether you think this is a cause for concern, as many do, or a good thing, as I do, it remains a fact that there is something that makes modern human civilization different from anything that has gone before.

So it's clear that something happened a few hundred years ago that changed forever the way that human life is organized. I would like to suggest that what happened was the discovery of a new and particularly effective way of discovering the nature of the physical world: the activity we call *science*. I would argue that this change, which reached something like its present state in the seventeenth century in Europe, is the single most important event in human history, particularly if you measure in terms of its impact on the lives of individuals.

Why study science? What else would you study if you wanted to understand the modern world?

When I want to make this point to my students, I ask them to perform a little mental exercise. I ask them to go back to England in 1775 and look for the thing that would have the biggest effect on human life in the next 2 centuries. Eighteenth-century England was a country riven by social strife, with problems of class and poverty that would make a modern American city look like Utopia in comparison. It was a country in the process of building the greatest empire the world had ever seen, a country whose most important colony was approaching a state of open revolt. And yet, if you wanted to see the future you wouldn't go to the Houses of Parliament or Buckingham Palace; you wouldn't even go to the universities and intellectual centers like

Oxford and Cambridge. Instead, you would go to a run-of-the-mill factory near the industrial city of Birmingham, to the firm of Boulton and Watt. There you would find a Scottish instrument maker by the name of James Watt working on a massive improvement of the steam engine, which could take the heat generated by burning coal and transform it into a powerful motive force.

When Watt was done, he had a machine that could actually use the energy that fell on the Earth millions of years ago in the form of sunlight and became trapped in coal. Liberated in Watt's boilers, this energy produced the steam that powered what we now call the Industrial Revolution. For the first time, we had a source of energy that didn't depend on human and animal muscle and wasn't tied to a specific place such as a waterfall. The steam engine ran the factories that drove the growth of cities; it ran the trains and ships that supplied those factories with raw materials and distributed their products throughout the world. It changed forever the relationship between human beings and the natural world.

So in Watt's England, the future was in an obscure technology, a technology that was probably invisible to most of Watt's contemporaries. I suspect that some professor 200 years from now will make similar statements about advances in genetic engineering in our own time, since these will be seen as being equally revolutionary. Just as the generation of the 1960s will forever be remembered for the first humans to walk on the moon, we will be remembered as the ones who first sequenced the human genome. This event will doubtless set into motion events that we can't even imagine today. Not to worry, though—I doubt if Watt had a clear picture of continent-spanning railroads or massive industrial cities like Manchester or Pittsburgh while he was tinkering with his engine, either.

WHAT SCIENCE IS: THE SCIENTIFIC METHOD

If science has had such an enormous impact on human history, it's natural to ask what, precisely, it is and how it differs from other human activities. Many authors (including me) have discussed something called the *scientific method* that is supposed to answer this question. I don't have any particular problem with this approach, provided that the authors realize that science, like any human activity, can't always be categorized into cookie-cutter organizational patterns. Rather than go through this entire exercise, however, let me instead lay out the two characteristics of the scientific approach to the universe that, taken together, explain why science is so different. These characteristics are observation and testing.

Observation

If you want to know about the world, you should look at it.and see how it works.

This statement is so obvious to us, living as we do in a secular, technological, twenty-first-century culture, that it's easy to forget that throughout most of human history it would have been considered, at the very least, wrong and quite possibly heretical. There have even been times when insisting on this proposition too vociferously might have gotten you burned at the stake. There have always been competing methods for discovering the true nature of the world, and it is, I believe, only the recent blazing success of science that has removed them from the intellectual forefront.

I can't resist giving one example of such a method, even though it is almost surely apocryphal. (I long ago learned, however, never to let mere truth or falsehood get in the way of a good story.) I heard it first from a fellow student at Oxford, a young man studying medieval history. It concerned one of those debates that was the staple of academic discourse in the Middle Ages. If you've ever heard of the great Scholastic debates, it's probably been in the context of the debate over how many angels could fit on the head of a pin. Don't sneer—this is actually an important theological issue, since it deals with the question of whether angels are corporeal or spiritual beings.

In any case, in my friend's story the debate wasn't about angels but about the number of teeth in a horse's mouth. Doctor A got up and quoted St. Augustine and the Church Fathers, while Doctor B relied on Aristotle and the Greek philosophers. As the debate wore on, a young man got up in the back of the room and said, "There's a horse outside. Why don't we just go look?" The manuscript went on, my friend assured me, to say that "they fell upon him, they smote him hip and thigh, and cast him from the company of educated men."

Whether the story is true or not, it illustrates a way of arriving at knowledge of the world that doesn't involve observation. In this case, the learned doctors were relying on received wisdom contained in revered texts. After all, they might have argued, what could you possibly see in the horse's mouth that Aristotle hadn't seen already?

Of course, you don't have to go to the Middle Ages to find this kind of attitude. The continuing battle in America about the teaching of creationism (and its current incarnation, Intelligent Design, about which more later) can be thought of as arising from the same source. On the one hand, you have scientists relying on observation of the fossil record and the evidence of DNA to understand how the Earth came to be the way it is. On the other, you have people who believe that the truth has been revealed in a specific text (in this case, the book of Genesis) and is not to be questioned.

You can find a similar attitude in people who populate that vaguely defined area we call New Age philosophy. When you meet their fuzzy insights about grand cosmic connectedness, harmonic convergences, flows of energy, or other such notions with arguments based on the laws of physics, you are often treated to a smug smile and the pronouncement, "There are some things that are not known to science." In this case, people arrive at the truth through something like an intuitive or mystical insight, and observation really has no role to play.

In general, I dislike the word *revolution* when it is applied to intellectual developments because it seems to imply a discontinuous break in what is usually a seamless progression of ideas. It makes a lot more sense to me to look for an event that can be taken to symbolize a development, rather than one that marks a discontinuous break. When I look for such an event to act as a symbol for the idea of learning about nature through observation, I go back to fifteenth-century Milan, where the Duke had just acquired a cannon, the prime piece of high-tech military hardware of its day. He called in his chief engineer, a man called Tartaglia, and asked him the kind of maddeningly simple question that people in government often ask their scientists: "How high do I point the barrel to get the cannonball to go the farthest?"

The point of the story is that Tartaglia didn't sit down and try to reason his way through the problem, as Plato might have recommended, nor did he consult revered texts, as the Doctors A and B might have done. Instead, he took the cannon to a field outside the city walls and started shooting it off, observing how far the cannonball went under different conditions. It would be a couple of centuries before Galileo and Newton would work out the laws that govern the flight of projectiles, but Tartaglia found—by trial, error, and observation—that if you point the barrel at about 45 degrees, you get maximum range.

Since that time, of course, the notion that scientific knowledge starts with observation has become a common piece of folk wisdom. Unfortunately, along with this understanding have come some misconceptions, and at the risk of belaboring a point, I'd like to clear them up. The most common misconception is that when scientists observe the world, they are supposed to do so with an open mind, with no preconceptions about what they are going to find. This idea often becomes a kind of straw man among philosophers who want to advance the notion of the social construction of the sciences.

Well, all I can say is that in my career—over 30 years in the scientific trenches—I have met only one scientist who actually operates this way. His name is Phillip Gingerich, a paleontologist and field geologist at the University of Michigan. Once when I was out in the Big Horn Basin of Wyoming with Phil, I asked him how he approached a new *locality* (that's the paleontologist's word for a place where they are looking for fossils). His response was revealing: "I go out into the field and let the rocks talk to me." Every other scientist I've ever known goes into an experiment, field study, or observation with a pretty clear idea of what he or she expects to find. In the words of philosophers, scientific observation is "theory laden." What makes good scientists isn't that they have no expectations at the start, but the fact that when the results don't match those expectations, the scientists believe what nature is saying. Instead of ignoring or suppressing the results, they change their idea about what those results should be.

In the historical development of the sciences, observations are typically followed by periods of intense intellectual activity during which regularities in the data are uncovered and used to develop the theories that characterize the scientific worldview. A canonical example of this sequence took place in astronomy in the sixteenth and seventeenth centuries. First, the Danish nobleman Tycho Brahe (1546–1601) spent his career accumulating the world's greatest collection of accurate data on the positions of the planets in the sky. (Astronomers don't like to be reminded of this, but such data had enormous economic value in those days because they could be used to cast more accurate horoscopes.)

When Brahe died, his assistant, a young German mathematician named Johannes Kepler (1571–1630), condensed the information in those reams of data into three laws, called Kepler's laws of planetary motion. One of them, for example, says that the planets move around the sun in elliptically shaped orbits, an insight that overthrew millennia of astronomical thought.

Finally, Isaac Newton (1642–1727) showed that Kepler's laws followed from a deeper theory that involved what we now call Newton's laws of motion and the law of universal gravitation. The end product of this chain of events was a profoundly compelling and profoundly simple picture of the universe, one in which you could see the motion of the planets as being like the hands of a clock, while the underlying gears represented Newton's laws. It was a universe in which the whims of the gods were replaced by the operation of rational, predictable laws—laws that could be discovered by the human mind. Scientists of the era often stated that their goal was to "discover what was in the mind of God" when he created the universe.

This so-called clockwork universe idea had a profound effect on the culture of the seventeenth and eighteenth centuries; you can see it in art, in literature, and in music. Indeed, some scholars have even argued that the framers of the American Constitution were deeply influenced by it; that is, they felt that they were discovering the laws that governed human society, just as Newton had discovered the laws that govern the solar system.

In the end, then, the process of observation leads to something very special. The human brain being what it is, when we find regularities in the world around us, we ascribe those regularities to a knowable cause, whether it be the actions of a spirit or the operation of a natural law. We produce a picture of the universe in which some cause produces the regularities we see. I call this process the creation of a theory, although I have to warn you that philosophers have conducted interminable debates on the question of what, exactly, a theory is. For our purposes, however, we can take Newton's laws of motion and the law of universal gravitation to be the paradigm of a theory of the universe.

Not all theories are created equal, however; some are much better at explaining what we see than others. Tycho Brahe, for example, produced a theory based on his measurements. In this theory the sun moved around the Earth, but all the other planets orbited the sun. In the end, his theory didn't explain the data very well, so it was abandoned. And it is this process of weeding out ideas based on how well they explain observations that gets us to the second important aspect of science—testing.

Testing

Many scientists would say that the most important aspect of the scientific process is the relentless comparison of the predictions of our theories to what actually happens in nature. In a sense, this gives science something that no other field of human endeavor has—an impartial (perhaps even brutally impartial) outside referee. It really makes no difference whether someone proposing a theory is a Nobel laureate or a newly minted postdoctoral fellow. If the predictions of his or her theory aren't confirmed by experiment or observation, the theory goes down in flames. Period.

It is this aspect of science that seems to be the most difficult for nonscientists to absorb, so I'd like to spend a little time on it. The problem is that it gets tied up with issues of impartiality and objectivity, issues that raise all kinds of warning flags in postmodern culture. So, without further ado, let me state my thesis in the bluntest possible terms: *In science there are right answers, and we know how to find them.*

Let me introduce this notion with another historical example. After Isaac Newton had enunciated the basic laws that governed the operation of the planets, his contemporary Edmund Halley decided to apply those laws to these mysterious, wandering lights in the sky known as *comets*. A comet, after all, is a light that suddenly, and unexpectedly, appears in the sky, moves around for a while, and then disappears—hardly the sort of thing to fit neatly into a clockwork universe. Using Newton's laws, Halley was able to determine the orbits of 24 historical comets and found that 3 of the comets had identical orbits. From there, it was a small step to the realization that this wasn't a situation with 3 separate comets, but of 1 comet coming back three times. Again using Newton's laws, he was able to work out when the comet would appear in the skies again. (Actually, this calculation isn't as simple as it sounds since the gravitational effects of the outer planets can produce significant perturbations in cometary orbits. In fact, Halley found that the comet's reappearance was delayed by almost 2 years due to these effects). In effect, he said, "If Newton's picture of the universe is correct, then at the end of 1758, this comet should once again be in the sky."

And sure enough, on Christmas Eve in 1758, 16 years after Halley's death, an amateur astronomer in Germany pointed his telescope toward the sky and saw the comet. To honor Halley, the comet was named after him. We now know of a Chinese sighting of the comet as far back as 240 BC, and the most recent (and rather unspectacular) visit was in 1986. (The next visit will be in 2061.)

This sequence of events, called the *recovery* of Halley's comet, completed the cycle of the scientific method. I have argued that science starts with observation, and in this example we see that it also ends with observation. It is precisely this interplay between human ideas and the implacable reality of nature that makes science different from other fields of intellectual endeavor.

One amusing sidelight to the Halley's comet story is a quote from Halley himself, written when he first calculated the comet's orbit: "Wherefore if it [the comet] should return around 1758, candid posterity will not refuse to acknowledge that this was first discovered by an Englishman!"[1] (Halley was, during his eventful life, both a sea captain and a secret agent for the British Admiralty. I like to tell people that he was the original James Bond.)

It is important to realize that it is possible to imagine a world in which the skies remained dark on that Christmas Eve in 1758. It could have turned out, in other words, that the predictions of the Newtonian theory would not have been verified, and that the theory would have been wrong. In the language of philosophers, this means that Newtonian physics is *falsifiable* (or, in the language of the law, *testable*). It is a theory that makes a prediction that could, in principle, turn out to be unfulfilled. It is now generally recognized that a theory that does not have this property is simply not a part of science.

It is also important to recognize that a theory can be falsifiable and still be wrong. The statement "The Earth is flat" is a perfectly testable and falsifiable statement. It turns out to be false. On the other hand, a theory can be falsifiable and correct. The statement "The disorder of a closed system cannot decrease over time," for example, is testable and falsifiable, but happens to be true (it's known as the second law of thermodynamics). That event in 1758 was one piece of evidence that Newton's clockwork universe is another example of a theory that is falsifiable but true.

The fact that scientific theories are testable played a major role in my own decision to become a physicist. In my undergraduate days I had a long infatuation with philosophy and seriously considered following a career in that field. But the more I read and the more I listened to the major philosophers of the day, the more I began to see that, although people in the field addressed the most fundamental issues, in the end the debates couldn't come to closure. As long as your arguments were logically consistent, and as long as you were as good a debater as the next guy, no one could prove you wrong. I pictured the world's philosophers as circling their wagons into their separate schools, with no outside agency capable of establishing a standard of validity. As I have progressed through my career, I have seen this same situation in other fields in the humanities. In exchange for the right to ask truly deep and fundamental questions, people in these fields must give up the possibility of coming to definite answers.

Science is different. There can be (and, indeed, often are) disputes over which theory best explains the available data—disputes that can go on for decades. In the end, though, everyone agrees that it is the data that will ultimately decide the issue. By restricting their attention to the types of questions that can be answered through observation, then, scientists are able to bring closure to their debates. At the same time, however, they have to give up asking the deepest questions (such as, What is the meaning of life?). In the words of the late Ken Boulding, "Science. . . [is] the process of substituting unimportant questions that can be answered for important ones that cannot."[2]

Having made this point about science in general, I want to point out one aspect of the process that has enormous implications for the way that the scientific enterprise interacts with society at large. The process of theory proposal and testing that I've outlined above works quite well, and eventually can be trusted to tease out the correct interpretation of nature. What it cannot do, however, is deliver those interpretations on a specific schedule, and this can sometimes cause problems.

Look at is this way: Every scientific idea starts out in the mind of a specific individual somewhere in the world. Over the course of time, that idea will be tested over and over again, and may eventually become accepted. It may even make it into textbooks. But there is inevitably a period of time when the idea is, as it were, in-process—still being tested, still on probation. The process of acceptance is a complicated one, involving individual choices by thousands of individuals in laboratories around the world. Between origination and acceptance, then, a scientific idea has to go through a kind of limbo when it is neither known to be wrong nor known to be right.

When we talk about the way that a citizen in the twenty-first century will encounter science, this indeterminacy becomes very important, because it contradicts one of the images that many people have of science. Many people think that science always has answers, and that those answers are always right. (If you doubt this, observe the way that the word *scientific* is used in everyday discourse.) Understanding that there are periods of uncertainty in the development of any idea is a hard one for many people to accept.

I have run into this attitude repeatedly when I have lectured on science to judges (something I do often under the auspices of the school of law at my university). I like to phrase the problem this way: "As a scientist, I am comfortable saying that the scientific method will produce reliable answers to a specific question after 10 years of research. The problem is that the trial starts next Tuesday."

The point is that many of the issues that people will encounter as citizens will involve science in the gray area between initiation and acceptance. Thus, learning about the great principles of science, though very important, is only part of what they need to know. They also need to understand how those ideas became accepted so that they can make reasonable judgments about current issues such as global warming or the latest warning about carcinogens in food.

Another feature of the scientific method is that, given enough time, it will ferret out both mistakes (which happen) and outright fraud (which also happens, albeit less frequently). When two scientists in Utah announced over the evening news several years ago that they had succeeded in producing cold fusion in their laboratory, they caused quite a stir. Speaking personally, it was a great time to be a physicist—people actually wanted to talk to you at parties! It only took the scientific community about a month, however, to reject the claims on the basis of technical flaws in the original experiment and the inability of other researchers to reproduce the results under controlled conditions. More recently, when a researcher in South Korea announced that his laboratory had produced human stem cells, inquiries triggered by reporters quickly uncovered the fact that the claim was fraudulent. Thus, although these sorts of events cause some commentators to question the very foundations of science, in fact they show that the system, left to itself and given enough time, will eventually come to the right answer.

Let me make one final point about the scientific method before I go on, because it involves another critique of the process that is sometimes raised by philosophers. The scientific method may be a way of obtaining closer and closer approximations to the truth, but it cannot—nor is it designed to—produce "Truth." The reason is simple: If the system is based on observation, then it is always logically possible that tomorrow some observation will be made that will over-throw some long-established principle.

Having said this, however, let me make an important distinction between what is logical and what is reasonable. It is not possible to prove logically that the sun will come up tomorrow; the mere fact that it has done so for 4.5 billion years in no way proves that it will do so tomorrow. On the other hand, we all recognize that it would be silly to organize our lives on the expectation that it won't. This is the difference between being logical and being reasonable. So, while it is logically possible that tomorrow someone will disprove the principles of the conservation of energy or evolution by natural selection, I would personally put the odds against that happening much higher than I would the odds of the sun not coming up tomorrow.

So that's the way things are. In an imperfect world, that's the best we can do, and, frankly, it seems pretty good to me.

WHAT SCIENCE ISN'T: PSEUDOSCIENCE AND ALL THAT

The successes of the scientific method have bred a kind of intellectual mimicry, a succession of things that try to look like science and claim to be science, but aren't. They go under the collective name of *pseudoscience*, and they often present a confusing problem for the casual observer. I'm talking about areas like astrology, extrasensory perception, the Bermuda Triangle, UFOs and extraterrestrial visitors to Earth, and the various versions of creationism. Fortunately, even the cursory description of the scientific method given above can give us tools to analyze and reject many of the claims made in the past as well as those that will—alas!—be made in the future.

Let's start with the first tenet of the scientific method—the notion that we learn about nature by observation. What this tells us is that the first thing we should do when confronted with what we think might be a pseudoscientific claim is to ask a simple question: "Are the facts, as presented, really true?" I'll give some examples of how this question can play out.

Astrology

Astrology is a belief system that goes back to the great Babylonian astronomers. A central tenet, simply stated, is that the position of the planets in the sky at the moment of your birth determines your personality and the events in your life. The most important aspect is what is called the individual's *sun sign*, which is defined, essentially, as the constellation in which the sun would appear to stand at the time of birth if we could see the stars during the daytime. The familiar signs, 12 in all, each covering about a month, are the constellations that mark a band in the sky called the zodiac, through which the sun moves in a year.

There are lots of ways to criticize astrology. You could point out, for example, that the regular movement of the Earth's axis has changed the relationship between the signs of the zodiac and birthdays since the time of the Babylonians. Today, there is almost a full month of slippage in these numbers, so your sun sign and where the sun actually is no longer align. You could also point out that modern astronomers assign 13 constellations to the zodiac instead of the astrological 12. The extra member, Ophiuchus—the serpent bearer—was added about 60 years ago, but never makes it into horoscopes.

But quibbling aside, the main question is, "Does astrology work?" Here's a little test I do occasionally with my students: I take yesterday's horoscopes from the newspaper and remove the signs, so that there what's left are twelve little bits of advice unidentified by birth date. I then ask each student to do two things: first, pick the advice that it would have been best to have had the day before, and then give his or her sign. In this way, you can see whether the horoscopes actually produced useful advice.

But here's the catch: If the students just picked advice at random, you would expect that they would get the right advice about one time out of twelve. The real question, then, isn't whether a particular individual got the right advice—some will, just by chance. The question is whether, taken over a large group, the results show that reading your horoscope does you any more good than picking advice at random. When I do this with classes of 80–200 students, the results are always consistent with the choice being random. Based on this simple test, astrology in its simplest form fails to meet the most fundamental test to which we can put an idea—the test of whether it actually works in the real world or not.

You normally don't have to go far to find other failures of astrology. I always present my wife and me as exhibit A in this category. We are both Virgos, born within 6 days of each other (although in different years). As we discovered on our first date, however, we not only have very different personalities, but we completely reverse the normal gender stereotypes. Wanda, a female musician, tends toward rationality and linear thought. I, a male theoretical physicist, tend toward the emotional and intuitive. Complementary strengths like this may work well in a marriage, but they sure don't do much for the believability of astrology when they show up in people with the same sign.

Finally, I should point out that astrology represents a prime example of what I call the golden age type of pseudoscience. In essence, it argues that there was wisdom gained in ancient times that has been lost to crass modernity and should therefore be respected because of its longevity. What can you say to something like this? I always ask people whether, if they had a toothache, they would go to a Babylonian dentist. It seems to work.

The Bermuda Triangle

Celebrated in myth and story, the Bermuda Triangle is supposed to be this place in the Atlantic where ships and airplanes mysteriously disappear without a trace. Again, the way to approach this problem is to ask whether more ships actually disappear in the triangle than in other stretches of ocean. Fortunately, the Bermuda Triangle also illustrates another interesting fact about pseudoscience. If a myth hangs around long enough, sooner or later someone will get so ticked off about it that they'll do the work of checking it out and write a book, thereby saving you the trouble of doing your own research.

In this case, a book titled *The Bermuda Triangle Mystery—Solved* by Larry Kusche, a research librarian and flight instructor, does the work for us. It tells the story of Kusche's seemingly interminable search through historical records, ending in a painstaking documentation of the building of the Bermuda Triangle myth by the media over the decades, watching reporters tailoring the facts to fit their stories. My favorite anecdotes are the ship that was supposed to have been lost in the Triangle but actually sank in the Pacific, and the "mysterious disappearance" of an aircraft that actually flew into a hurricane.

Ancient Astronauts

I believe that every scientist, sometime in his or her life, owes it to the community to put in some time debunking the pseudoscience du jour. This is a discouraging task, something like taking out the garbage; no matter how much you clean up today, there'll always be more tomorrow.

I did my time in the late 1970s, when ancient astronauts were all the rage. You still see this sort of stuff on late night TV, but it seems to have faded from the general consciousness. The basic premise of the theory is that people like the ancient Egyptians were too stupid to have built something like the pyramids by themselves, and so must have been helped by an obliging group of extraterrestrial visitors. A sample proof of this conjecture, found in many books, is the claim that the pyramids are laid out in a perfect square and the Egyptians couldn't have done anything so precise.

So, following my check-the-facts dictum, I spent a rainy spring afternoon in the University of Virginia library looking up actual surveys of the pyramids. (You could probably do this online these days.) I found that far from being a perfect square, the shape of the base of the Great Pyramid of Cheops is actually 8 inches longer on its longest side than on its shortest. Given that the length of a side is almost 1,000 feet, this is a phenomenal job of surveying for guys stretching rope over empty sand, as the Egyptian engineers did. But how good would it be for surveyors today?

To get an answer I went to the guy who taught Civil Engineering 101—the standard beginning engineering course that sends generation after generation of students out to measure the main quad of their campuses. "If you sent your students out to survey a square and they came back with this," I asked, showing him the pyramid survey, "what grade would they get?"

He answered, "They'd flunk."

So there it is. To believe in ancient astronauts, you have to believe in a race of extraterrestrials with the technology to design interstellar spacecraft, but who would flunk Engineering 101.

So if debunking pseudoscience through fact checking is so easy, why is pseudoscience still around? There are many answers to this question. One involves the need of some people to believe in something greater than themselves—in this case, in a race of benevolent extraterrestrials who will help us and, perhaps, even watch over us. Another is the fact that there are very few professional rewards for scientists who spend their time in this way. Even the simple checking of one claim embedded in an entire book of such claims took me a couple of days, and from the point of view of my career I would probably have been better off had I spent those days working on my research papers. So don't be surprised if legitimate scientists look at pseudoscience, shrug, and get back to the laboratory.

Creationism

Sometimes pseudoscience becomes something more than a minor and amusing irritation and starts to present a real danger to American education. The continuing efforts to push some form of creationism into the public school curriculum is a good example of this phenomenon. For my purposes, I will define *creationism* as the belief system that holds that the origin of the universe as described in the book of Genesis is to be taken as literally true, that the universe and all living things were created in pretty much their present form, and that the creation was the act of God (or, in recent times, an Intelligent Designer).

There is a long history of creationist activism in this country, and a long string of court cases that have turned back successive attacks on the biology curriculum. These cases usually involve the First Amendment to the Constitution, which states in part that "Congress shall make no law respecting an establishment of religion. . . ." The string of court cases starts with the Scopes trial in 1926, when Kentucky tried to outlaw the teaching of evolution, and is followed by *MacLean v. Arkansas* (1982), when Arkansas tried to mandate "balanced treatment" of evolution and creationism; *Edwards v. Aguillar* (1987), when Louisiana tried to say that the two should be taught together or not at all; and, most recently, by *Kitzmiller v. Dover Area*

School Board (2005), which banned the reading of creationist statements in classrooms.

In most of these cases, the arguments centered around the interpretation of the First Amendment, but the scientific issues also played a role. Thus the cases give us a fascinating insight into the evolution (if you'll pardon the term) of creationist thinking over the past century. And this, in turn, gives us a chance to see how different aspects of the scientific enterprise have to be brought into play.

In the beginning, creationism was fairly unsophisticated: The Earth and all living things were put in place by God and have changed not at all (or very little) since then. In one popular version, called Young Earth Creationism, this event took place about 6,000 years ago. If you wanted to get really precise, some creationists argued that creation took place on Sunday, October 23, 4004 BC at 9:00 in the morning, Greenwich Mean Time. (Geology students regularly throw parties on October 23 to celebrate the foundation of their field.)

But this quickly led to problems, because there is ample evidence that the Earth is older than that. For example, we can see stars billions of light-years away, even though the light from those stars wouldn't have had time to reach us if everything started 6,000 years ago. To explain how the Earth could be so young but appear so old, the doctrine of Created Antiquity was developed. Actually, the first book on this subject, *Omphalos* by Philip Henry Gosse, was published 2 years before Darwin's *The Origin of Species* in 1859. *Omphalos* in the Greek means "navel," and the argument was that Adam was created with a navel, even though he had never been inside a womb. In a similar manner, the creationists argued, the universe was created to look *as if* it were much older than it actually is. Those distant galaxies, for example, were created with light already most of the way to Earth.

This doctrine certainly explains all the data, so I can't attack it as I did the Bermuda Triangle and ancient astronauts. The rejection of this classical version of creationism, by scientists as well as by the courts, hinged on another feature of science discussed earlier in this chapter—falsifiability. The fact of the matter is that there is no experiment or observation that could ever prove the doctrine false. Any new finding would be met by the bland assurance, "Well, the universe was just created that way." Far from being a strength, this feature of early creationism shows that it is simply not a part of science, because every scientific theory has to be falsifiable. Thus the courts were right to find it to be a thinly disguised religious teaching and ban it from the public schools.

More recently, after a brief period of something called Creation Science, a new version of creationism, called Intelligent Design (ID), took center stage. The central thesis of this new approach was a concept called *irreducible complexity*. In essence, the argument is that there are systems in nature that require many different parts, but which will not function if one part is missing. The flagellum of a bacterium, the clotting of blood, and the immune system were all put forward as examples of irreducible complexity. Since the probability that all the parts of these systems would come together at the same time by the working of natural selection is negligible, the argument goes, they are proof that living systems were created by an Intelligent Designer (ID advocates never use the word *God*). Intelligent Design had its day in court (6 weeks, actually) in the town of Dover, Pennsylvania. The school board had voted to have a statement about ID read in the ninth-grade science classes. Teachers refused to read it on the grounds that the statement violated their obligation to teach the best science; board members came into class to read the statement; parents sued; and the whole thing wound up in federal court.

This is not the place to rehash all the scientific evidence for evolution and against ID, although if you have the time, Judge Jack Jones's 139-page decision in the case is a real masterpiece.[3] His key finding, after hearing all of the scientific testimony, was that Intelligent Design is not science at all, but a thinly disguised religious doctrine. Scientists testifying for the plaintiffs basically demolished the ID arguments in a careful point-by-point refutation of the claim that natural systems exhibit irreducible complexity.

Based on my discussion of the scientific method in this chapter, what are we to make of Intelligent Design? For what it's worth, here is my analysis. The statement "There are systems in nature that exhibit irreducible complexity" is a perfectly good scientific hypothesis that can be tested by observation. In this sense, ID avoids the trap into which the Young Earth Creationists fell. On the other hand, when the statement is tested, it turns out to be false. Thus ID is an example of a theory that is falsifiable but false, like the theory that the Earth is flat. Thus the debate comes down to whether you want to say that ID is *not* science or is *bad* science. Whichever option you prefer, ID clearly doesn't belong in the public school curriculum.

CONCLUSION

In the end, then, there are a relatively small number of characteristics that define the enterprise we call science. The central ideas involve observation of the world and the constant testing of theories against nature, with the requirement that everything that is to be called science must be testable.

In retrospect, this seems like a relatively modest set of requirements for an intellectual endeavor that has wrought such incredible changes in the human condition. How this simple method has revolutionized human existence, producing new forms of wealth and well-being in the process, will be the subject of Chapter 7. Before I get to that, however, we will turn to the question of how this method produces a body of knowledge with which every citizen needs to be familiar. In other words, I will turn to the topic of scientific literacy.

CHAPTER 2

Scientific Literacy: What Is It?

Much of the rest of this book will be taken up with a discussion of how best to produce scientific literacy in our educational system, so it behooves us to have a clear definition of what it is before I start to talk about implementing programs. I'm going to get to scientific literacy in a somewhat roundabout way, because the context of my definition is at least as important as the definition itself. I see scientific literacy as one piece of a much larger construct called *cultural literacy,* a piece with its own special issues and difficulties. What I am going to do, then, is to start with the wider topic, then focus down on that segment of cultural literacy that pertains to science and technology and can properly be labeled *scientific literacy.* In later chapters I will discuss the reasons that scientific literacy is important and suggest ways of incorporating it into our educational system.

CULTURAL LITERACY

Cultural literacy is the knowledge that educated people, at a given time and in a given place, assume that other people possess.

For example, in tomorrow's newspaper there may be an article about Mexico City. The writer and editor do not say, "In Mexico City today—and oh, by the way, Mexico City is the capital of Mexico, which is a country that shares the southern border of the United States." They don't say this because they assume that their readers know where Mexico City is, where Mexico is, and something about current and past relations between Mexico and the United States. And because they assume that their readers already possess this knowledge, they don't bother to tell readers what it is. And that, in turn, means that if the readers don't already have that knowledge, they are just out of luck. They won't know what the article is about.

The sum of all of the bits and pieces of information that people assume we posses is what is called cultural literacy. It's important to understand that what I am talking about here is purely a matter of empirical fact. Like it or not, these assumptions get made, and someone who is missing important pieces of this body of information will not be able to understand parts (perhaps important parts) of the discussion that goes on in society at large.

When I want to make the point about people making these sorts of assumptions in real life, I often use cartoons from some publication like *The New Yorker*. My favorite cartoon shows two bearded men, dressed in long robes, surrounded by pairs of animals. All of them are standing on a large wooden boat that is clearly in the process of sinking. The tagline for the cartoon is "I knew those woodpeckers were going to be a problem!"

Okay, it's a funny cartoon, but I'd like you to think for a moment about all of the knowledge that the cartoonist assumed you had. He didn't say anything about Noah, about the book of Genesis, about 40 days and 40 nights of rain, or even about the animals two by two—he assumed you knew all that. The point is that if you didn't know it, you'd have no idea what the cartoon was about. And although understanding a cartoon may seem to be trivial, *every* communication, trivial or serious, has the same kind of assumptions built into it. And if the communication is about something like a contemporary political issue instead of a cartoon, not getting it may have serious consequences both for the individual and for society at large.

To emphasize the fact that I am advancing an empirical claim, let me tell you about an experience I had a number of years ago. For many years I was a science consultant and contributor to *Smithsonian Magazine*, a publication that reaches several million well-educated Americans. One day I was having lunch with members of the editorial staff and I decided to try a little experiment.

"Would you," I asked, "use the word *impressionism* without explaining it to your readers?"

"Oh, sure—they'd know what that was."

"Okay—how about *abstract expressionism?*"

"Oh no—we'd have to explain that."

There was, in other words, a clear dividing line in their minds between what their readers already knew and what would have to be explained. Since that time I have run through this same exercise with all sorts of people involved in communicating with the public—book publishers, writers, broadcasters, and so on. Every one of them has the same kind of dividing line in their minds. Part of their stock-in-trade is, to use my terminology, a clear understanding of the boundaries of cultural literacy.

Of course, those boundaries may vary somewhat from one kind of publication to the next, but they don't vary all that much. One of my favorite ways of illustrating this point is to refer to an article that appeared in the sports pages of *USA Today*—hardly the place where most of us would expect high levels of cultural literacy from the readers. The article concerned a brawl at a hockey game, and began with the sentence "When these guys meet on the ice, it's not like Socrates and Plato talking under a shade tree." That was it. No mention of Greece or philosophy or the golden age of Athens. The writer and his editors simply assumed that the reference would call up images of bearded, toga-clad philosophers peacefully exploring the intricacies of some esoteric subject, and the "oomph" of the sentence came from the contrast with the guys trading punches out on the ice.

My own introduction to the concept of cultural literacy came about in a strange way. I was sitting in my office at the University of Virginia when I got a phone call from a colleague, E. D. Hirsch. Don was a professor of English and was already well along in developing the basic theoretical concepts for the field. At that point in time he wanted to move on to spelling out its contents explicitly. He had gone to the dean to see about getting support for the project, and the dean had asked a simple question, "Do you have any scientists involved?" That question led to the phone call, and as Don explained his ideas, my initial reaction was, "Hey, that's a neat project. We should be able to finish it off over a couple of beers some evening." Now, over 20 years of development and elaboration later, I sometimes wonder what I would have done had I known how the project would grow.

In any case, Don enlisted Joe Kett, a prominent historian, and the three of us set out to answer a simple question: If there is, indeed, a body of knowledge that represents American cultural literacy in the (then) 1980s, what knowledge, precisely, is contained in that body? It is, after all, one thing to argue in the abstract for something like the concept of cultural literacy, quite another to get explicit, point to something, and claim that it is something that everyone needs to know.

The first thing we realized was that cultural literacy was not just a collection of facts (a frequent complaint brought against us). It is a potpourri of things—facts, ideas, connections, images. I have come to think of it as a kind of complex matrix of knowledge, a matrix into which new ideas and concepts can be integrated.

I have also begun to think about cultural literacy (and the subfield of scientific literacy) as something like a building code for the educational system. If you look at the building code in your community, you will find long lists of rules—if you want this size window in this wall, then this size beam has to be over it; if you want an electrical outlet in this location, then it has to be wired this way; and so on. The point is that the building code defines a minimum standard for structures, and nothing can be built in the community that falls below that standard.

In the same way, I would argue, no one should leave the American education system without acquiring the matrix of knowledge defined by cultural literacy. But just as people can (and often do) put up structures that go beyond the local building code, there is no reason that education has to stop with filling in the cultural literacy matrix. I would expect, especially at the university level, that higher level skills would continue to be taught. In this scheme, then, cultural literacy becomes the bedrock on which the rest of a person's education rests.

I like to illustrate the importance of cultural literacy with an example from Hirsch. Think about the following text:

> First you put things into piles. There may be more than one, depending on how much there is to do. Then you have to find the equipment. Once you have that, you're pretty well set.

If you're the way I was when I first encountered this text, you are probably having a feeling of mental floating—what in the world is he talking about? The whole thing seems to make no sense. Now let me supply you with the matrix of knowledge you need to understand this text: Imagine a heading like "Doing the Laundry" above the text. The "click" you just felt probably does more to establish the importance of cultural literacy than any argument I could make.

As it turns out, there is a great deal of research in the psychology of learning to back up the idea that people learn best when they are able to fit new information into an existing matrix of knowledge. In fact, it is this characteristic of the human mind that Don Hirsch likes to call up when he reminisces about his discovery of cultural literacy. He was investigating the role of what we might call "good grammar" by giving selected groups of students texts written in good and bad English, then testing them for speed and comprehension. As you might expect, he was finding that students did much better with texts written in good English. At a community college in Richmond, Virginia, however, he was shocked to find that none of the students did very well with the texts, regardless of the quality of the English in them. The text involved an interaction between Stonewall Jackson and Jefferson Davis, and when he interviewed the students, he found that they didn't know who these men were! In terms of the example above, they didn't know about the laundry, and this made it difficult for them to understand the passage they were asked to read.

When Hirsch's book *Cultural Literacy* was published in 1987, it included a long discussion of the widespread literature supporting the notion that people with well-filled-in matrices of knowledge do better at learning than those whose matrices are less well fleshed out. More controversially, the book also included a list of terms that represented, in our minds, the minimum knowledge needed to function in American society. The subsequent *Dictionary of Cultural Literacy* (by Hirsch, Kett, and Trefil) expanded the list, providing definitions and cross-connections. And despite the fact that we had to approach some 20 publishers before we found one courageous enough to take us on, both books have done quite well since then. I am always amazed (and gratified) when I walk into someone's office and see the *Dictionary*, often well-thumbed, sitting on a shelf.

Once our arguments became public, there were serious objections raised to the entire cultural literacy enterprise in the academic press, as one would expect. I've run into these objections so often since then that I routinely deal with them in my lectures, rather than wait for them to come up in the question period. I give here the three main objections and my responses to them:

Isn't cultural literacy just about memorizing facts?

As I pointed out earlier, there is a lot more than facts involved in the matrix of knowledge called cultural literacy. It does involve facts, of course, but is also involves concepts, generalizations, principles, connections, and laws. But suppose this weren't the case; suppose it was just a matter of memorizing a lot of facts. Would this be so terrible?

When I was growing up in a blue-collar neighborhood in Chicago, the kids used to play a game called alley ball. It was played with a baseball bat and tennis balls. The idea was that each kid would pick a National League team to represent (I grew up in a part of the city where you inherited the Cubs and the National League), and when you went to bat, you would go down the batting order of the team you were representing, batting either right- or left-handed according to the handedness of the player involved. Thus every kid was expected to know the batting order of all the National League teams and whether each player was right- or left-handed. Memorizing these lists wasn't considered particularly onerous; it was just part of living in the neighborhood, and it pretty much happened by osmosis.

The point is that children will naturally memorize all sorts of miscellaneous stuff—batting orders, the shapes of cars, the lives of movie stars, and so on. What would be wrong with tapping into this skill and having them memorize something useful? Provided the process doesn't descend into an exercise in rote learning, a little memorization never hurt anyone.

And for the record, I can't recall a single time in my career when I needed to know the batting order of any National League team, either in the 1950s or later.

Doesn't this ignore the multicultural nature of American society?

Let me use another example from my childhood to deal with this point. I grew up in an ethnic neighborhood, where you were as likely to hear Czech being spoken on the streets as English. During the day, I went to a public school and there, in English, I learned about what I have called American cultural literacy. A couple of afternoons a week, however, I would go to a different building in a different part of town, and, in a different language, learn about the Czech and Moravian culture that was my personal heritage.

This experience gave me a clear view of multicultural America. I picture the country as being like a collection of houses built around a large common. Each house has its own backyard, and in some backyards, like mine, the language might not even be English. That backyard is where you go to be with people who are like you. If, however, you want to talk to people in the other backyards—people who aren't like you—then you have to go out into the common. This is the area where we communicate with every American, and this kind of communication requires a shared base of knowledge. It is this knowledge that I have called cultural literacy, and in this sense cultural literacy is all about the common and only peripherally about the backyards.

If you think about a multicultural society like the one outlined above, you quickly realize that without the glue provided by common cultural literacy, it would fall apart. There would simply be no way for people in different backyards to talk to each other. My sense is that over the last several decades educators and scholars have tended to focus their attention on the backyards and ignore the commons. Paying attention to cultural and scientific literacy is, in effect, an attempt to undo some of this neglect.

Doesn't cultural literacy change over time and from place to place?

Yes. This issue is usually raised to argue for the impossibility of actually defining the content of cultural literacy, even if you accept the fact that it exists. Fortunately, on this issue there is now an accumulation of data that can lead us to quantitative answers to questions about temporal and geographical changes. The *Dictionary of Cultural Literacy* was first published in 1988, and is now in its third edition. The second edition, published in 1993, was driven by the massive political changes attendant on the collapse of the Soviet empire. It suddenly became important, for example, to know where Kazakhstan is. The third edition, published in 2002, was driven primarily by the explosion in information technology and molecular biology of the 1990s. This meant that terms like *clone, stem cell,* and *web page* had to be added.

Because of this history, we have a long record of the evolution of cultural literacy. Basically, it seems to change at a rate of somewhat less than 1% per year—about the same rate as changes accumulate in the English language itself. Thus keeping current with cultural literacy is no more difficult than tracking the inevitable changes in an ordinary dictionary.

Incidentally, the experience of going through these revisions gave me some insight into the first objection raised—the one that raises the "fact" issue. I was amazed to discover that a number of entries in the science sections had to be revised because of events in the former Soviet empire. For example, in the astronomy entry for *satellite*, there was a cross-reference to the *satellite nations* of Eastern and Central Europe. (For the benefit of younger readers, this is the group of countries we now refer to as the New Europe.) The fact that we had to delete this cross-reference is a good example of the interconnected nature of knowledge in the cultural literacy matrix.

As far as cultural literacy varying from one country to another goes, there is also data from the *Dictionary* that can be brought to bear because the work has been translated into a number of foreign languages. In this process, publishers typically put together a panel of scholars to make the text relevant to their readers. The results are pretty much what you would expect. The science sections remain largely unchanged (science is, after all, no respecter of national borders), while the literary and historical sections are somewhat modified. The Swedish edition of the *Dictionary*, for example, has an entry for *Gustavus Adolphus,* but none for *John Adams.*

So in the end, I can say with some confidence that cultural literacy does, indeed, change over time and from place to place, but it changes in predictable and manageable ways.

SCIENTIFIC LITERACY

One way of looking at scientific literacy is to say that it is the portion of cultural literacy relating to science and technology. When we do this, we have to recognize that there is a subtle difference between these two kinds of literacy. The argument for cultural literacy rests on the notion that citizens need to know certain things because other people will assume they know them. When we turn to scientific literacy, this argument changes somewhat. We are still talking about knowledge that people need to have, but in this case the reason for the need is different. When we talk about what pieces of science to include in our matrix, we have to ask what knowledge people need to have to participate in the kinds of debates that will involve them as citizens. The need to know is still the basic argument, but the reason for that need shifts subtly when we move from the humanities and the social sciences to the natural sciences. In the former case, the argument is based on the empirical fact that people will make assumptions about what you know. In the latter, it has to do with the kind of issues that are likely to arise in an individual's life.

Given this fact, we can start to put together a definition of scientific literacy. What people need to know about the universe we live in are its basic operating principles. In my own teaching, I often characterize my attempt to inculcate these principles as supplying students with a set of mental DVDs, each explaining some aspect of the world around us. When a student runs across a news item about stem cells, for example, he or she pulls out the DVD labeled "cells" and looks at the section on development and gene expression. The student then has the background needed to understand what the news item is about. In effect, the student has been given the "Doing the Laundry" cue that we used to illustrate the concept of cultural literacy above.

This leads to a general definition of scientific literacy:

> *Scientific literacy* is the matrix of knowledge needed to understand enough about the physical universe to deal with issues that come across our horizon, in the news or elsewhere.

Let me amplify this definition with an example. As I write this, the debate about stem cell research is heating up again. As I will point out in Chapter 3, this debate isn't primarily or even mostly about science, but rather involves the way people look at the value of human life and the question of when the fetus or embryo acquires that value. The point, however, is that if you don't know what a stem cell is or understand why they might need to be derived from embryos, you simply can't get into the debate. In effect, you are like someone reading the text I gave earlier and not knowing about the laundry. Pick any other current debate—global warming, endangered species, the use of pesticides—and you can specify some scientific background that is needed before a person can participate meaningfully in the national dialogue on the subject.

Of course, as I will show in the next few chapters, there are other reasons one might want to have this matrix in place, but this will serve as a preliminary definition. The actual content of scientific literacy (what one might call its *operational definition*) can be defined in a number of ways. In Chapter 12, for example, I lay out a scheme for education in scientific literacy based on the great principles that tie the scientific world view together. Alternatively, one can turn to the *Dictionary of Cultural Literacy,* which contains a detailed list of concepts related to scientific literacy that takes up 134 pages. At this point, however, we needn't descend to that level of specificity.

When we move from a theoretical definition like this to the practical problems involved in teaching scientific literacy, we find ourselves in a classic "good news, bad news" situation. The good news is that, unlike in other disciplines, there is very little disagreement among scientists about what forms the core of their discipline. You would have to work pretty hard, for example, to find someone who would argue that Newton's laws of motion or the laws of thermodynamics aren't a central part of understanding the universe. In this respect, science educators have an advantage over people in fields like English literature, where there can be endless battles over which novels belong in the canon and which do not.

But there is also bad news, for the fact of the matter is that as poorly as other areas in the curriculum have done in producing a culturally literate citizenry, the sciences have done much worse. Let me call on another of my *Smithsonian* experiences to make this point. As I mentioned, the magazine reaches an upscale, educated audience. In spite of that, I would never have dreamt of using a word like *proton* in *Smithsonian Magazine* without explaining it. I would have to write "the proton, one of the particles that makes up the nucleus of the atom. . . ."

I call this device the *fatal descriptive clause,* because it signals to the readers my belief that they don't know what the word means and therefore they need my help to understand what's going on. One way of knowing whether an author thinks a particular term is a part of cultural literacy, in fact, is to see whether the author routinely attaches a fatal descriptive clause to it. (For the record, my wife informs me that the technical term for this sort of phrase is *appositive.*)

The poor state of public knowledge of science has important implications for the definition of scientific literacy, because it requires that we be a little more prescriptive here than we would be, for example, in discussing English literature. If scientific literacy is to be defined in terms of what people need to know to understand their world, then it's not going to be a simple catalogue of things that can be characterized as common knowledge. It will be, instead, the product of the deliberation of the best scientists and educators on the question of what knowledge people actually need to confront science and technology as they impinge on an individual's life.

Just as objections were raised by academics against the whole concept of cultural literacy, an approach to science education that emphasizes scientific literacy is going to rub many scientists the wrong way. In Chapter 10 I will discuss the different kinds of goals that have been enunciated for general education in science and present my case for scientific literacy in detail. In this opening discussion, however, I will confine myself to pointing out a few things that scientific literacy *isn't*.

It isn't about math.

The natural language of science is mathematics. It's not a historical accident that our first modern scientific view of the universe had to wait until Isaac Newton and Gottfried Leibniz had invented the calculus. It probably doesn't surprise you that I, as a theoretical physicist, would take this view, but even a field paleontologist, with a sweaty hatband and dirt under his or her fingernails, will use sophisticated mathematical techniques to analyze data. In the end, science is quantitative, and the natural language in which to express quantification is mathematics. That's all there is to it.

This does not mean, however, that people need to master the mathematical formulation of the sciences to be scientifically literate. One of the best-kept secrets in the world is the fact that the basic ideas of science are mostly pretty simple, and can be understood by almost anyone without recourse to full-blown mathematical statements. The general idea (though not necessarily the full quantitative rigor) can usually be captured in a simple sentence. Take Newton's second law of motion as an example. In mathematical form it is $F = ma$ (force equals mass times acceleration). I would argue that the statement

"The harder you push on something, the faster it goes" is a good enough rendering of this law for the purposes of scientific literacy. You really don't need the math.

To make this point, I often talk about what I think I'm doing when I write about science for the general public. In my mind, what I do is take a question that is asked in English, translate it into mathematics, use the rigidly defined rules of mathematical "grammar" to find the answer, and, finally, translate that answer back into English. By going through this process, I remove the requirement that my reader be able to do the mathematics in order to be able to understand the basic points about the science.

Incidentally, this analogy turns out to be very useful when you have to explain why you aren't able to answer questions that can't be translated into mathematics but make sense in English. "What came before the big bang?" is a good (and frequently asked) question of this type. The inability to answer has to do with the fact that "before the big bang" is not a concept that can be translated into mathematics. (The question is equivalent to asking, "What's north of the North Pole?" The problem isn't that there is nothing north of the North Pole; it's that there's not *even* nothing north of the North Pole.)

Some people object to avoiding mathematics in this way, on the grounds that to really understand science you need to include math. You could, however, make an analogous argument that to really understand Tolstoy you need to read him in Russian, and to really understand Homer you need to read him in classical Greek. These statements might be true, but are they any reason to deny students *War and Peace* or *The Iliad* in translation? I think not, and I would argue that there is no reason to deny students an understanding of the universe, even if that understanding is translated from mathematics to English. To do otherwise would constitute a false kind of rigor and do them a serious disservice.

It's not about doing science.

In the same way, the idea that the only proper goal of a science education is the ability to do science at some level is, in my mind, misguided. I will discuss my problems with this notion in more detail later, but for the moment let me argue by analogy with another field—the performing and studio arts.

When I was a student, some of the most useful courses I took had names like Music Appreciation and Introduction to Art and Architecture. These courses added immeasurably to my life, not only by giving me a deeper appreciation of their subjects, but by reinforcing the idea that music and art were meant to be a part of my life, despite the fact that my career was in science. Even though I came from a home where classical music was valued, these courses built on those childhood experiences, giving my experiences a kind of intellectual framework that they would not otherwise have had. And that, to be sure, was the whole point of this kind of course offering.

The most interesting thing is that never once, in the many years that I participated in these sorts of courses, did anyone ask me to play a musical instrument, compose a piece, or execute a painting. The instructors realized that their goal was to help students like me deepen their experience of music and art, not produce it. I would suggest that the same attitude would be useful in thinking about scientific literacy. As I will argue later, science is just as capable of deepening and enriching our lives as are music and art. It makes no more sense to demand that students learn to do science in order to gain this appreciation than it does to demand that they learn to play the violin before they attend a symphony concert.

It's not about technical competence.

When science education comes up in public discussion, particularly in a political setting, one of the most common statements that ussd to get made was "No one knows how to program a VCR." This statement is supposed to illustrate how scientifically illiterate the American population is.

I beg to differ. The ability to program a VCR or fix a car or figure out what all the buttons on a remote control are may be useful in the modern age, but it has nothing to do with scientific literacy. Scientific literacy has to do with an understanding of the structure of the universe in which we live and the ability to apply that knowledge in one's life. It may or may not be accompanied by the technical skill needed to operate electronic devices. Like mathematics, technical skills constitute a separate body of knowledge from scientific literacy.

In addition, I would argue that it makes no sense to try to incorporate these kinds of skills into the school curriculum as part of science.

We don't teach courses in Telephone Literacy for the simple reason that learning to operate a telephone is something that is acquired outside of the curriculum. We can (and do) get children started on computers, but I sense that even computer operation is becoming such a universal skill that we will soon be able to assume (with some obvious exceptions) that students are as familiar with computers as they are with phones. But other than a general introduction to computing, teaching the kinds of specific skills involved in programming a VCR would do little good for the simple reason that the pace of change in electronics is so rapid that anything students learned today would be obsolete by the time they graduated. For these reasons, I would argue that technical competence, while important, does not deserve the attention implicit in being part of the curriculum.

And for the record, I never did learn how to program my VCR.

CHAPTER 3

Scientific Literacy: The Argument from Civics

Pick up a newspaper on any given day and you are likely to find a story about a public issue that involves science in some way. Today, those stories may be about global warming, our energy future, and the use of stem cells, but no one can possibly predict what they will be 10 years from now. We can be certain, however, that whatever topics preoccupy people 10, 20, or 30 years down the road, at least some of them will require some basic understanding of science. And this, in a nutshell, is what I call the argument from civics as a justification for scientific literacy.

Consider the current debate over stem cells as a prototype of the problems of the future. In the United States, this debate has unfortunately gotten entangled with what is probably the most contentious and unresolvable issue on our political landscape— abortion. My personal views on this matter are laid out in a book I wrote with my friend and colleague Harold Morowitz, titled *The Facts of Life*. For our discussion of stem cells, however, it doesn't matter if you are pro-life or pro-choice. In order to make a connection between your views on abortion and the stem cell debate, you have to understand what stem cells are and how they are derived. If I were laying out the matrix of knowledge required to understand the science behind the debate, I would include the following understandings:

1. As cells in an embryo divide, they become specialized and are no longer able to turn into any kind of adult cell.
2. Up to about 8 cell divisions, cells do retain the ability to develop into any adult cell (a property called *totipotence*), and hence are called stem cells.

3. The most promising way to obtain stem cells is to harvest them from an embryo, killing the embryo in the process.

I would argue that this is the minimum background that would allow you to enter the debate. Once you have it, you can go on to the nonscientific parts of the discussion. If you are pro-life and believe that a single fertilized cell is a human person who should be protected by law, then the act of harvesting stem cells is, essentially, murder. If you do not believe that a collection of a few hundred cells with human DNA has the equivalent moral value of an adult or newborn person, then the medical benefits to be gained from the research—real benefits for existing human beings—far outweigh any harm done by killing the embryo and the potential life it represents.

Both of these positions can be taken by sincere, intelligent people. Neither allows room for compromise—either you harvest the stem cells or you don't. This is why I called the issue irresolvable. But the point is that unless you understand something about basic developmental biology, you can't even get into the debate. You have no intelligent way of bringing your moral sensibilities to bear, simply because you don't understand what is being debated.

There are many ways of defining the word *democracy*, but for me a good working definition is to say that in a democratic system, people who are affected by a decision have a say in how that decision gets made. In the United States, that say is largely exercised through the political process, and the attendant debate takes place in the print and broadcast media. A person who has not been equipped with the matrix of knowledge we call scientific literacy will be excluded from large areas of that debate, and simply will not be able to make his or her voice heard.

In fact, if you want to make gloomy predictions, it's not hard to imagine the country evolving along either one of two equally undemocratic lines. On the one hand, things may be perceived as being so complex that only a technological elite would be allowed to make decisions. (Oddly enough, I seldom hear this course urged by scientists.) The other, perhaps more frightening, option would be for rational debate to be abandoned altogether and for people to follow leaders who are essentially demagogues. Either one of these paths would be a disaster as far as maintaining the United States as a democratic society is concerned, and the surest way to avoid either outcome is to see that citizens are equipped with the knowledge they need to make their own decisions.

EMERGING TECHNOLOGIES AND PUBLIC DEBATE

This isn't going to be easy. In the future, a good proportion of the political debate in this country is going to be generated by advances in scientific and technological fields that are just coming over the horizon now. Stem cells represent the tip of the iceberg as far as the emerging field of biotechnology is concerned, but fields like information technology and nanotechnology will not be far behind. Let me make a short digression here to speculate about some of the issues that we might be confronting a decade from now, just to give you a flavor of the complexity that scientific advances are likely to bring to our political lives.

In the nineteenth century we learned one of the great central truths about living systems, that they are based on chemistry. When we say that something is alive, in other words, we do not mean that it possesses some mysterious "vital force," but simply that it is running a particular set of chemical reactions. In the twentieth century, we moved on to understand that life's chemistry is intimately connected with a molecule called DNA, and we are now busily working out the details—the gears and sprockets, if you will—of the chemistry of living systems. We are, in a very real sense, learning how to get under the hood of these systems and manipulate them for our own purposes. And, as the case of stem cell research shows, this new ability will introduce an entirely new set of problems with which citizens will have to grapple.

Take cloning as an example. As Ian Wilmut demonstrated when he produced Dolly the Sheep in 1996, it is possible to remove the DNA from an egg, replace it with DNA from an adult animal, and have the resulting cell grow into a fully mature individual. The countless "Hello Dolly" headlines reflected the fact that, although this process had been known to work on amphibianssuch as frogs, Dolly was the first clone of a mammal. And since sheep are uncomfortably close to humans on the tree of life, this raised all sorts of potential problems.

The most obvious (and, ironically, the least interesting scientifically) is the prospect of producing a human clone. Headline after headline called up visions of marching armies of identical clones—a nightmare vision if there ever was one. Even in the sports pages, there was speculation about putting together basketball teams of Michael Jordan clones. On the darker side, there were scenarios of clones being raised to supply transplant organs for their DNA donors, a particularly gruesome take on the old "insane cousin chained in the basement" fairy tale.

Unfortunately for the tabloids, these kinds of speculation ignore one of the basic facts that we know about human development. It may have come as a shock to psychologists and sociologists, but it is true that over the last decade we have come to realize that a fair amount of human behavior is determined by our genes. No person is born as a tabula rasa—something any parent could have told the researchers. On the other hand, no person is born as a machine, doomed to play out the instructions in his or her genes. Working out the complex interplay of genes and environment that produces an adult human is, and will remain for some time, a major area of research. Right now, if I had to guess, I would say that the split will turn out to be something like 50-50. The point, though, is that just because two people have identical genes does not imply that they will be identical as adults—think of twins as an example. What made Michael Jordan a champion was partly his physical endowment, but more important than that was the determination and dedication he developed over his playing career. A clone with a different history could easily turn out differently; one wag commented that a Michael Jordan clone could easily turn out to be a violinist instead of a basketball player!

From the point of view of scientific literacy, however, the important point about the cloning story is that we do not have a word in the English language for a human egg whose normal DNA has been replaced by the DNA of another adult. And as any linguist can tell you, if you don't have a word for something, you are going to have a hard time dealing with it at the moral, legal, religious, or political level.

At the moment, scientists are virtually unanimous in saying that, with current technology, human cloning is much too risky to be morally acceptable. Fair enough. I have to tell you, though, that I sometimes wonder about this statement. Personally, I can't imagine why people would want to produce a clone of themselves; but there are enough people in the world with outsize egos so that I'm sure there are individuals out there who would like nothing better. The actual cloning procedure isn't that hard to do—a small building, a few million dollars' worth of equipment, a government willing to look the other way, and some women willing to donate eggs are all you really need. If someone wanted a clone badly enough and was willing to ignore the moral aspects of the issue, there is little doubt in my mind that it could be done somewhere in the world. In fact, I have a fantasy that someday I'll be sitting in a meeting where gray-bearded savants are discussing the ethics of human cloning when a young man will walk into the room and announce that he is just such a clone. Then what?

A more serious debate will take place over something called *therapeutic cloning.* In this technique, which is still in its very early stage of development, someone who needs a new organ donates DNA to produce stem cells. Using the burgeoning technique of a field called *tissue engineering,* scientists might then be able to grow that organ so that surgeons could implant it. Because the cells in the new organ would have the same DNA as the patient, the patient's immune system would not reject it, thereby eliminating one of the greatest risks attendant on organ transplant procedures today. This is the dream of a new field called *regenerative medicine.*

At the moment, this scenario is very much in the future, but it's not hard to imagine that it will be turned into reality in a decade or two. At that point, an entirely new set of moral issues will arrive. If I need a new heart, how do we weigh the moral value of saving my life against the moral value of the embryo that has to be killed to do it? What if we fertilize several eggs and don't use all of them? Are we morally compelled to find surrogate mothers for the others? Store them forever? Flush them down the toilet? You can't imagine people wrestling with these kinds of issues if they don't know some basic molecular biology.

Incidentally, William Haseltine, one of the pioneers of regenerative medicine, has pointed out that if the field succeeds in fulfilling its promise, some very simple questions are going to be very hard to answer. What do you say when someone asks, "How old are you?" when the answer varies from organ to organ?

And what if we learn enough about genetics to manipulate the DNA of an embryo to produce some desired result? What if, for example, we engineer an athlete's DNA so that he produces his own steroids? Would that athlete be allowed to compete in the Olympics? What if the new gene was the result of a natural mutation instead of genetic engineering? You won't even be able to escape molecular biology on the sports page!

Once we really get going on the mechanics of life, there's no end to the problems we'll face. Most of us would have no problem dealing with someone whose kidneys have been grown from stem cells and transplanted. Such a person would clearly be *human*, no matter how you define the term. But what if the transplanted organ included computer components? What if the transplant was done to make the person smarter or more efficient, rather than to save his or her life? What if more than one organ was changed? At what point does that person stop being human and start being something else? Transhuman or even (God forbid!) superhuman? Have we defined the concept of *humanness* well enough to even begin dealing with these sorts of questions?

These questions may sound like science fiction now, but I have heard every one of them raised by well-known scientists at dignified, mainstream scientific conferences. There is no question that we or our children are going to have to deal with at least some of them in the foreseeable future. Fulfilling your obligations as a citizen isn't going to be easy in the future, and fulfilling them without scientific literacy is going to be impossible.

Similar issues are going to arise with advances in information technology. Believe it or not, I have already heard discussions among scientists about whether turning off a computer in the future will be a morally permissible act. As computers become more and more capable, the question of whether they are intelligent and conscious gets more difficult to answer. I personally do not believe that computers can replace the human brain, a conclusion dealt with at length in my book *Are We Unique?* Nevertheless, it's clear that at some point in the future computers will be much closer to recognizable consciousness than they are now. When they reach that state (think HAL in the film *2001: A Space Odyssey*), what will be the moral judgment that has to be made if we want to turn them off? Would that be equivalent to murder? What if we want to send them on a one-way mission to a distant planet or star system? These are not easy questions to answer, and if it becomes a political issue (imagine an organization called People for the Ethical Treatment of Computers), it is going to require a great deal of sophisticated thinking on the part of the electorate to resolve. Again, it's hard to see how this could be done without high levels of scientific literacy.

I can't resist giving one more example, even though this one is really far out. One of the most interesting advances of the 1990s involved a technique called *quantum teleportation*. Basically, this is a process that uses the methods of quantum mechanics to destroy a *photon*, a particle of light, in one location and create an identical photon somewhere else. Think of it as a primitive version of the transporter on the television series *Star Trek*, a fictional device that works by destroying a person's atoms in one place and assembling identical atoms somewhere else. Scientists in Austria already have used quantum teleportation to send photographs over optical cables through distances of several miles, but imagine going forward to the point where you can send people through a quantum transporter. Is the person who comes out the other end, identical atom-for-atom with the person who went in, really the same person, whatever that means? As one of my colleagues asked, only half in jest, would that person have to pay your income taxes?

REAL ISSUES INVOLVE MORE THAN SCIENCE

This little diversion into science fiction serves to emphasize the fact that complex issues await us in the future, issues that will require high levels of scientific literacy if ordinary citizens are to have some control over their futures. But they also illustrate two other important points about the place of scientific literacy (and science in general) in public debate:

1. Important issues in our society never involve questions of science and technology alone.
2. The amount and type of scientific knowledge needed to function as a citizen is limited.

The second of these points is well illustrated in the above discussion of stem cell research and cloning. To get into these debates you have to understand the science involved in producing organs from stem cells, and clones. You don't have to understand a lot; it's not necessary that you be able to sequence a strand of DNA to understand therapeutic cloning, for example. What we see instead is that there is a minimal matrix of knowledge necessary to get into any debate that will arise around these issues, and that once you are past that minimum matrix the serious issues revolve around things that are only tangentially related to science.

We can, in fact, identify three distinct aspects of this kind of debate, corresponding roughly to questions of fact, of values, and of policy. The first of these centers around issues of whether it is possible to get the benefit of stem cell research without destroying embryos—by finding ways of manipulating other types of cells, for example. This is a scientific question, and understanding the debate requires some level of scientific literacy. Once we are past that level, however, we are into an area that has very little to do with science. As pointed out above, all of us, once we understand what the debate is about (questions of fact), have to find a way of applying our own moral values to the situation (questions of value) and deciding what to do about it (questions of policy). In the case of stem cells, you can imagine individuals advocating anything from massive government research support to criminalization of the entire field.

The same point can be made about the deeper issues attendant on advances in the understanding of the mechanics of life. Once again, the science serves as an entry point into the debate, but the real questions involve something else. If, for example, you are going to confront the questions of what a human being is, and where you draw the line between a human being and something else, the science can only take you so far. Similarly, if you are going to deal with the very serious issue of the moral value to be assigned to a developing human being as it progresses from zygote to embryo to fetus, you won't be able to do it by science alone.

Let me elaborate on this statement for a moment, because it is a good illustration of the point I am trying to make. Basically, the abortion issue (and all of the other issues relating to it) comes down to the distinction between the terms *human being* and *person*. *Human being* is a biological term, referring to an organism of the species *Homo sapiens*. It is a term that can be defined in terms of physical structure or of DNA sequences. Human being, then, is a term that can be defined totally in scientific terms.

Person is another matter. Person is a legal term, and denotes someone who is entitled to legal protection. Personhood is defined differently from one society to the next. In our society, for example, it is definitely conferred at birth, and the abortion debate is mainly about whether this conferral should happen earlier. There are other societies, however, where personhood comes some time after birth. In medieval Japan, for example, a human became a person at the first cry, and if a midwife killed a deformed child before this happened it was not considered an act of murder. Many non-Western societies, in fact, considered deformed babies to be "ghost children," and were horrified when Christian missionaries told them that killing them amounted to murder. The decision as to when a human being becomes a person, then, is not something that can be given a scientific definition, but has to be recognized as a question of societal values rather than of fact.

In traditional Christian theology, the problem of the conferral of personhood was tied up with the process of *ensoulment*. The issue was framed in terms of the question of when the developing fetus acquires a soul. Thomas Aquinas (1225–1274), basing his argument on the appearance of aborted fetuses, placed this time at 40 days for men and 90 days for women (don't ask!). In 1869, Pope Pius IX, speaking ex cathedra at the First Vatican Council, in effect, declared that ensoulment occurs at the moment of conception, thereby setting the stage for the current abortion controversy. In both cases, however much the argument started out in scientific terms (When does a fetus start to look human?), in the end there is a leap from humanity to personhood, and that leap has nothing to do with science.

You can see this same shift from scientific to nonscientific arguments operating in almost any debate that has a scientific or technological component. Take the ongoing debate over the use of nuclear power as an example. Discussions about nuclear power touch many subjects that are the domain of science and technology: the nature of radiation, the production of nuclear fuel, and the generation of electricity from high-pressure steam. This means that citizens need to know something about how a nuclear reactor works, which parts a reactor has in common with other generators, and which are different. They should know something about radioactivity, such as what it is, why it can be dangerous, and what steps are necessary to shield humans from it. These sorts of facts and concepts constitute the matrix of knowledge that forms the background to the entire debate.

But with this knowledge in place, the real issues can surface. Once all the technical nit-picking was done in the 1970s, for example, the nuclear power debate came down to a simple question: How much risk are you willing to take in order to get cheap electricity? That particular debate got settled on the side of risk aversion, and we have not licensed a new reactor in this country in decades. Although I believe personally that that was the wrong choice, the question was settled by and large by democratic choice.

Actually, this particular debate seems to be heating up again, this time driven by the awareness of the possibility of global warming and the role played in that threat by the burning of fossil fuels like coal. The new debate will require some extra pieces of scientific literacy, relating this time around to climate dynamics and environmental change, but the essential point will be the same. Once we enter into the debate and have acquired the minimal scientific background needed to understand the issues, individual choices will be made according to the value we assign to the environment as opposed to the value of relatively cheap energy. Is it more important, for example, to protect people against whatever risks may arise from the building of new reactors or to protect the environment from the addition of carbon dioxide to the atmosphere? Once again, the debate is not entirely about science, but involves issues that all of us must resolve according to our own moral calculus.

We could go on giving examples, but I think the point is clear. For each debate in which a citizen is likely to be involved, it is possible to define a minimal matrix of knowledge needed to understand what the debate is about. I think of this knowledge as a kind of entry pass into the civic arena. Once inside, however, the debate shifts, and the scientific issues take a back seat to less quantifiable, though arguably more important, issues of personal values.

This pattern, as I show in Chapter 10, has important implications for the kind of science education we provide for our students, since it will allow us to come up with an approximate definition of the content of scientific literacy. For our purposes in this chapter, however, I want to stress the minimal nature of the scientific knowledge needed to function as a citizen, and to do this I want to let you in on a dirty little secret in the scientific community.

There is a common mythology to the effect that when scientists enter into the civic debate on an issue, they do so from a position of superior knowledge of the technical aspects of the problems being discussed. This is often (though not always) true of the people who appear as talking heads and represent particular scientific points of view to the public. It is not, however, true of the scientific community in general. Unless the subject happens to fall within an individual's particular sphere of specialized research competence, chances are that he or she doesn't know a whole lot more about it than the average informed citizen. Thus part of the secret is that scientists are usually pretty much in the same boat as everyone else when it comes to the matrix of knowledge needed to participate in public debates.

I have to tell you, this realization came as quite a shock to me. It was in the 1970s, when I was a newly minted (and rather naive) PhD. The subject at issue in public discussion was whether the United States should follow the lead of France and Britain in developing a commercial supersonic airliner. This was a period when the idea that one ought to think about environmental impacts of new technologies was just coming into play, and a group of scientists started to sound the alarm about possible damage to the upper atmosphere from a constant influx of jet exhaust at high altitude. (One of the features of the proposed aircraft, as I recall, was that it would fly at altitudes well above those attained by normal planes to cut down on wind resistance.)

At the time, I was one of a small group of theoretical physicists exploring and developing the idea that things called *quarks* were the fundamental constituents of the matter of the universe. (The idea was eventually accepted, although today things called *strings* have been proposed to be the fundamental constituents of quarks—*sic transit gloria mundi!*) I was definitely a scientist at the time, but I hadn't a clue as to what the debate was about.

The reason is simple. If you want to talk about the effects of pollution in the upper atmosphere, you have to know (1) what chemicals are being introduced, and (2) what will happen to them in that environment. Answering the first question is easy; today, you could probably just look the answer up on Google. The second question, though, is complicated because it involves chemical reactions of unusual elements in a very thin, cold gas being flooded with ultraviolet radiation from the sun, with perhaps some ice crystals thrown in just to make things tricky. And the debate, of course, centered around what would happen if you threw lots of those chemicals into the stew.

Like most physicists, my knowledge of chemical reactions went back to courses I took as an undergraduate. Nothing in my own research was remotely connected to the questions at issue in this particular debate. Other than a sort of generalized knowledge of chemical reactions, then, my scientific training had given me no more information about the subject of this debate than might be expected of the average college graduate.

So what did I do? That's where the dirty little secret comes in. I looked at the people involved in the debate, asking friends and colleagues about their general reputation within the scientific community. As it happened, one of the people on the negative side of the debate was the late Hans Bethe, a Nobel laureate who was one of the giants of modern physics as well as a frequent advisor to the government on matters scientific. He was a man deeply respected among scientists, a feeling with which I, based on a few brief encounters, heartily concurred. So when he talked, I listened. I assumed that he had gotten numbers right, and decided that the argument against the supersonic plane had more merit than the argument for it.

Actually, in the end it was the economic arguments—arguments to which I paid little attention at the time—that were the most important aspects of the debate. When the supersonic Concorde finally flew, it was a financial disaster, and was eventually shut down by the sponsoring governments. But if we confine our attention solely to the scientific part of the debate, the point I want to make is that there is no magic knowledge possessed by scientists that gives them some sort of special status on civic issues. If they are honest about it, most scientists will be able to give you stories similar to mine. Except for the small number of people whose research happens to fall squarely in the middle of a contested subject, scientists need the same kind of generalized knowledge that everyone else needs when they enter into their function as citizens.

But the fact that the scientific knowledge needed to enter public debate is minimal and the fact that science is seldom the sole factor influencing public choice doesn't change the central argument, which is that without this knowledge, no one can make reasonable choices about these issues. Carl Sagan put it this way in his book *The Demon-Haunted World:*

> We've arranged a global civilization in which the most crucial elements. . . profoundly depend on science and technology. We've also arranged things so that no one understands science and technology. This is a prescription for disaster. [1]

CHAPTER 4

Scientific Literacy: The Argument from Culture

Is science really part of culture? I suppose the answer to this question depends on what you mean by the word *culture*. In the fullest sense, when culture is understood to be the social and physical web in which all human beings live, there is no doubt that science would have to be included. After all, as I pointed out in Chapter 1, it is precisely the ability to understand and manipulate the natural world that distinguishes *Homo sapiens* from other species.

But the word *culture* is universally used in a more restricted way, in the sense of what we might call *high culture*. I'll define this as the body of knowledge that someone in a given society needs to master in order to be accepted into the company of educated people. Thus in twenty-first-century America someone who had never heard of Shakespeare or Mozart would be considered uneducated, regardless of the amount of money he or she had. In the vernacular, that person would lack class.

Writing that paragraph calls to mind a snippet from a TV comedy show I saw years ago. A comedian playing a Prohibition era gangster had just been brushed off by a woman because he was "uncouth." Later, the gangster asks his cronies what that means. "It means you ain't got no couth, boss." Peeling some bills off a fat wad, the gangster says, "Well, go into town and buy me some. Take the truck."

One way of getting at this notion of culture, then, is to ask what someone in modern America has to be able to talk about intelligently to keep from being labeled *uncouth*. There is little doubt that the arts—painting and music, for example—would fall into this category. So would a certain subclass of writing that generally goes under the name of *literature*.

As someone who has written many books, I am always bemused when I look over the roster of authors invited to speak at things like book fairs and literary festivals. From those rosters, you would conclude that literature consisted almost entirely of fiction writing and poetry. I almost never see a nonfiction author on the list, and certainly not anyone writing about science. Without wanting to seem paranoid, I have built up an impression over the years that science is simply not welcome (or, for that matter, even considered much) in circles of people who consider themselves educated. One of the great attractions of the cultural literacy project for me, in fact, was the self-conscious inclusion of science as a part of that very culture.

And in the end, I suppose that this is the crux of the argument in this chapter. It is a truism to say that modern Western society has been shaped by the development of science over the last 3 centuries. Including a knowledge of science in the intellectual armamentarium of educated people in our society would, therefore, seem to be virtually automatic—what my students call a "no brainer." The fact that science is not usually regarded this way is one of those things that people tend to accept as a matter of course, but when you do think about it, it begins to seem strange. So the first thing we have to ask is how a dominant force in our society—some would say *the* dominant force—came to be relegated to a dark corner of the educational mansion. In what follows I will look at two important historical phases of thinking on this issue, then talk about what education in scientific literacy might do to make science more central to culture.

C. P. SNOW AND THE TWO CULTURES

It's impossible to get very far into a discussion of science and culture without referring to the seminal work that, for better or worse, has defined the terms of debate for the past half century. In 1959, the British chemist, novelist, and government official Charles Percy Snow delivered the prestigious Rede Lecture at Cambridge University. The lecture was titled *The Two Cultures and the Scientific Revolution* and was later published as a book of the same name. His basic thesis was that the scientific and literary cultures (in England, at least) had grown to be mutually exclusive, and he took the literary types to task for remaining willfully ignorant of one of the most important forces shaping their society. In what has become the most widely quoted passage in the book, he put it this way:

> A good many times I have been present at gatherings of people who, by the standards of the traditional culture, are thought highly educated and who have with considerable gusto been expressing their incredulity at the illiteracy of scientists. Once or twice I have been provoked and have asked the company how many of them could describe the Second Law of Thermodynamics. The response was cold; it was also negative. Yet I was asking something which is about the scientific equivalent of "Have you read a work of Shakespeare's?" [1]

Needless to say, this argument touched off a furious debate among English literary types, including a vicious ad hominem attack from F. R. Leavis, one of the country's leading humanists. Among his least venomous comments was the charge that Snow was "intellectually as undistinguished as it is possible to be." [2] (Rereading this screed a half century later, though, I noticed that Leavis never indicated that he had any inkling of what the Second Law of Thermodynamics is about.)

Just to level the playing field before we go on, let me explain that the most commonly cited statement of the second law is that an isolated system cannot become more ordered over time. Think of a series of time-lapse photographs of a teenager's bedroom and you pretty much have it. The second law has enormous consequences in our lives—it decrees, for example, that fully two thirds of the energy in coal burned to generate electricity must be lost as waste heat to the atmosphere—and forms one of the central pillars of classical science. (For anyone wanting a readable but in-depth discussion of the topic, I highly recommend Hans von Baeyer's marvelous *Maxwell's Demon: Why Warmth Disperses and Time Passes.*)

Having dealt with the second law, we can return to Snow's central thesis, that practitioners of the sciences and the humanities have somehow walled themselves off from each other. I suspect that this phenomenon was (and remains) more common in British intellectual circles than American. Nevertheless, I have encountered hostility to science often enough at faculty meetings and conferences to know that many of the attitudes Snow criticizes are still around.

Let me take on my scientific colleagues first. In most American universities students of science and engineering are required to take courses outside of their field—the so-called area requirements—to receive their degrees. In some institutions some thought has been given to the courses the students take, but more commonly the science faculty simply looks the other way and lets the students adopt a Chinese menu approach to their nonscience education. At meetings where thoughtful science faculty members discuss the curriculum, I have often thrown a simple question onto the table: "What training would you like our graduates to have outside of their field of specialization?"

The answers almost invariably come back in this order: (1) We would like them to be better at communicating, particularly in writing; (2) we would like them to have a better appreciation of the role that science and technology play (and have played) in society; and (3) we would like them to have an overview of all the sciences so that they can understand how their field fits into everything else. (I have, incidentally, gotten an almost identical list when I have raised this question with scientists from private industry.)

Once the discussion with my faculty colleagues has gone on for a while, I drop the other shoe. "Okay, if that's what you want, how about talking to the English department about introducing a 3-semester-hour course devoted to writing for science majors?"

At this point you can just see the people getting off the train. "Oh no, there just isn't any room in the curriculum."

Sometimes, just to be provocative, I start pushing. Couldn't this or that technical course be dropped or deferred to graduate school? How about lowering the required number of elective science and math courses to make room for writing or a science and society course? I am sorry to report that in an entire career in academe, and in countless hours spent in curriculum committees, I have never been able to push the discussion past this point.

My colleagues are not being obtuse. They have a clear goal in mind: They want to produce the very best scientists and engineers that they can. Consequently, they focus on cramming the maximum amount of technical knowledge into the curriculum. This has, as the above discussion shows, the unfortunate effect of limiting the time these students can spend acquiring skills outside of their specialty.

This is the reason, I suspect, for the standard stereotype of the nerdy scientist, uninterested in anything but his research (the stereotype always seems to involve a "he"). The highly specialized curriculum actually does science students a serious disservice, because very few of them will spend their entire careers at the lab bench or computer screen.

Eventually, they will move up and start interacting with management and the public, and at that point skills like communication become extremely valuable. In addition, this blinkered approach to science and engineering education helps to create one of the strangest phenomena I know of—the fact, outlined in Chapter 3, that working scientists comprise one of the most scientifically illiterate groups in our society.

Having said all this let me make another observation. Despite these kinds of curricular restrictions, it has been my experience that scientists participate in the literary and artistic cultures more than most groups. Go to a concert, play, or opera at any university and you will see a good proportion of the science faculty in the audience; in fact, my general impression is that you are likely to see a higher proportion of people from science departments than from English or philosophy. Somehow—perhaps because of all those music appreciation courses they took as undergraduates—many scientists seem to have acquired a taste for the arts.

There is a certain folklore that bolsters this claim. There is a legendary connection between theoretical physics (certainly one of the most left-brained activities around) and music (which operates largely in the other cerebral hemisphere). The icon for this connection might be Albert Einstein's violin, but I've seen enough of it during my career to know that Einstein was far from an exception. During the folk dance boom of the 1970s and 1980s, when I was a dance instructor and performance director, I routinely found that almost all of my male dancers were scientists, mathematicians, or computer jocks. I guess there's a kind of relief that comes from shifting from a highly mental activity to a physical one.

But there is another reason why scientists find it relatively easy to cross the divide between the Two Cultures. In a word, it's language. As I pointed out earlier, the natural language for the sciences is mathematics, a highly specialized language that requires a fair amount of training to use. In America, on the other hand, the natural language of the humanities is English, something we all speak from childhood. This means that when a scientist wants to get to know a play of Shakespeare (to return to C. P. Snow), all he or she has to do is pick up the script or (preferably) attend a performance. This may not give that person an in-depth understanding of the play, but, then, Snow wasn't asking for an in-depth understanding of the second law, either—just a simple description.

I think this natural linguistic difference explains the existence of a strange double standard dividing the Two Cultures. We all know that everyone at an intellectual dinner party will be expected to talk intelligently about a new novel or film, but not about the latest results from the Hubble Space Telescope or string theory. I have to say that, over late-night drinks, I have often heard many of my scientific colleagues express the same kind of irritation with this situation that I feel. It just seems unfair. However, I also have to say that absent the kind of reforms I propose in Chapter 12, I don't see the situation changing any time soon.

What about the other side of the Two Cultures divide? Let me start by noting that just as many scientists participate in the arts, many of my colleagues in the humanities and social scientists are perfectly conversant with the latest scientific developments. Besides being the creator of the concept of cultural literacy, for example, Don Hirsch is also one of the most scientifically and technically savvy people I've ever known, despite the fact that he is a professor of English, and my George Mason colleague James Pfiffner, a political scientist who specializes in the study of presidential transitions, constantly sends me clippings to keep me abreast of developments in my own field. In addition, I have encountered little of the overweening superciliousness from literary types that so irritated Snow a half century ago. This may be due to the fact that I've spent most of my career in the United States.

Nevertheless, I think Snow's basic point—that people on the humanistic side of the Two Cultures gap are largely ignorant of science—is as valid today as it was in 1959. You can see one reason for this by looking at the science part of the area requirements at American universities—the flip side of the requirement that scientists and engineers take courses outside of their specialties. These requirements, which haven't changed since Snow's time, almost always require that the student take a year of science courses. Typically, there is little monitoring of the content of those courses, so that in some institutions a course like a History of the American Environmental Movement can satisfy the requirement as well as an actual science course.

Now I have nothing against teaching the history of the environmental movement. It's just that a course like this should not constitute half of a college education in science, any more than a specialized course like theater lighting, valuable as it may be, should be allowed to be half of a scientist's exposure to the humanities.

The consequence of this lack of exposure of humanists to the sciences plays out in strange ways in American intellectual life. For example, when we were putting together our list of items that constitute American cultural literacy while assembling *The Dictionary of Cultural Literacy*, we sent copies out to many of our colleagues (both friends and critics). A large percentage of these people were well-known public intellectuals—historians, educators, political commentators, and so on—and many of them sent back long, thoughtful analyses of our work. What struck me on reading these replies was that, while the scientists offered comments across the board, other people would go into great detail about the literature and political sections, then beg off with a sentence like "but of course, I'm really not qualified to talk about the science." The fact that these prominent and well-educated people lacked the confidence to analyze science in the same way that they analyzed every other aspect of American culture is as good a piece of evidence for the persistence of the Two Cultures gap into modern times as anything I know.

THE TWO CULTURES TODAY: THE POSTMODERN TURN

Academe being what it is, there has also been a new spin put on the Two Cultures gap over the last couple of decades—a spin that Snow would never have dreamt of. I'm referring to something that developed into a phenomenon called the Science Wars of the late 1990s.

If you've never heard of these wars, you're in good company—most scientists haven't, either. (I have often wondered, in fact, whether you can have a war when one side isn't aware that it's going on.) Basically, the Science Wars grew out of the bizarre postmodern philosophy that infected the humanities departments of American universities in the late twentieth century (and is now, I am told, on its way out). Deriving from French literary theory, this point of view emphasized the social construction of knowledge, usually denying the validity of the notion of objective truths. Its most extreme proponents (and there were plenty of those) descended into a kind of sophomoric solipsism, using the sort of arguments that led Samuel Johnson to kick a rock and say, "Thus I refute [Bishop Berkeley]."[3]

In Chapter 1, I pointed out that science is different from other intellectual activities because it possesses an impartial arbiter of ideas—

nature herself. To a scientist, once experiment has spoken, the argument is over. It was precisely this aspect of science—what I would argue is the core of the whole enterprise—that the postmodern thinkers attacked.

Before going further, I should say that there is no question in my mind that there is, indeed, a social component to science. Scientists talk to each other, and their ideas are developed in a social context. Furthermore, science is embedded in the larger culture and shares the prevalent preconceptions and modes of thought. Isaac Newton, for example, could no more have conceived of the principles of relativity than he could have written rap music. In addition, social and political processes can speed up or slow down the advancement of a particular area of science; think about the current American situation regarding stem cells as an example. There is, then, no question that the general social atmosphere has an effect on science.

Because scientists form their own community, there are also internal constraints on research within each discipline, constraints that often have little to do with the requirements of the science itself. In the early twenty-first century, for example, it became almost impossible for scientists who were skeptical of the global warming orthodoxy to get their work published in major journals. The general attitude was that expressing doubt about widely publicized predictions was playing into the hands of politicians who wanted to ignore the warnings of mainstream researchers. And while I personally find this kind of behavior on the part of editors disgraceful, in the long run it tends to backfire. One of the great joys of the scientific life is to see a major journal forced to change its policies in the face of overwhelming data. In the end the editors are scientists, and like all scientists they will eventually accept, however grudgingly, the verdict of the natural world.

I remember an incident in the 1980s that illustrates for me the limits on social control on the sciences. This was the time when the notion that the dinosaurs were driven to extinction by the impact of a large asteroid was just being put forward. Today this is a textbook idea, but at the time many geologists found it profoundly disturbing. The reason for their unease was primarily historical. The foundational struggles of their science came from the attempts of early geologists to champion the notion of the gradual evolution of the Earth's surface, an idea that was strongly opposed by proponents of catastrophic events like Noah's Flood. Having won that historical battle through Herculean efforts in the eighteenth century, geologists weren't about to allow anyone—particularly physicists—to slip another kind of catastrophe into the planet's history.

Because of this philosophical mindset, there was a lot of to-ing and fro-ing about publishing work related to asteroid impact. I can remember, for example, *Smithsonian Magazine* being forced to pull an article on the controversy because a senior Smithsonian paleontologist objected to the magazine publicizing something he compared to the Bermuda Triangle.

In any case, during the height of this controversy I was at a seminar given by a sociologist of science. She presented a detailed study of the political infighting involved in promoting the asteroid hypothesis—who talked to which editor, who tried to block publication of what paper, and so on. Her analysis was a good example of the old adage about making sausages: This birth of a scientific idea often isn't a pretty thing to watch.

None of the scientists in the room was particularly surprised by her analysis. After all, we are fully aware that our community has its own brand of office politics. What shocked us, though, was that the speaker seemed to think that it was the politicking that determined the validity of the asteroid hypothesis. Our disbelief mounted until finally one senior paleontologist couldn't take it any longer. He got up and asked, "Is it news to sociologists that data matters?"

You see, to the scientists in the room, the politics were an inessential detail that really had very little to do with the asteroid hypothesis. What mattered to us was the accumulation of data that supported it, because we knew that in the end, data wins. To our postmodern colleague, however, it was the data that was an inessential (albeit inconvenient) aspect of the process, and she clearly believed that what led the community to accept the hypothesis were all those behind-the-scenes phone calls to editors and the like.

So this particular wrinkle on the Two Cultures comes down to a question of what different communities think is important. In general, scientists don't object to people studying the internal workings of our community, although to tell the truth most of us find it a little boring. People who are fascinated by office politics, after all, are unlikely to choose scientific careers. But what matters to us in the end is data. We insist that there is a fundamental difference between something like the Second Law of Thermodynamics and the social convention that a red traffic light means *stop*.

People on the other side of the Science Wars, at least those at the extremes, often deny that there is a real difference between the two. To them, all science is social construction, like traffic laws. It is the total frustration that scientists feel with this viewpoint that caused one physicist to invite his opponent in a debate to test the social construction of gravity by stepping off a balcony of the high-rise building where the debate was being held.

I am indebted to Brian Boyd, professor of English at the University of Auckland, for a brilliant analysis of this problem in his 2006 article in *The American Scholar*. He sets out the postmodern agenda in terms of two postulates: (1) *antifoundationalism*, the doctrine that nothing, including science, can provide an unassailable basis for truth, and (2) *difference*, the doctrine that all statements of truth grow from local standards, and hence cannot be universal. (The latter is roughly what I call social construction.)

Leaving aside the logical issue of claiming, as a universal truth, that there can be no valid claim for universal truth, we can get to the heart of the matter. If you go back to my explanation of the scientific method in Chapter 1, you will find at its very core that science is about the constant testing of ideas. There are not now, and never have been, any universal truths in science, just better and better approximations to reality. Antifoundationalism is so much a part of science that we never even talk about it anymore. In fact, Boyd tasks the postmoderns with "discovering" something about science that has been an integral part of the endeavor for a long time. He traces this notion back to Darwin, and I would push it even farther back, certainly to Newton, and perhaps earlier.

When I read Boyd's article, it suddenly dawned on me why I felt such high levels of irritation listening to the smug pronouncements of the postmoderns in various colloquia and conferences. It's the same feeling we get when a pimply adolescent goes on as if he had just discovered sex for the first time in human history. Science does not produce eternal and unchallengeable truths? As my students would say, "Well, duh!" You don't need eternal truth to produce profound changes in the human condition or a vastly expanded understanding of the universe in which we live. Our current approximation seems to do pretty well in that regard, thank you.

In general, as I indicated above, the Science Wars have had very little impact on the sciences themselves—a strange situation when you think about it. I have always thought it somewhat bizarre that there are recognized fields of academic endeavor with names like sociology of science and philosophy of science, but that these fields have no impact whatsoever on the operation of science itself, for the simple reason that few scientists pay much attention to them.

There are many reasons for this, but one is that those of us who have actually looked into the postmodern literature have come away pretty unimpressed with what we saw. All too often we find analyses that are driven by a generally leftist political agenda, with data chosen to strengthen preconceived positions. For example, several sociological studies of high-energy physics experimental groups have been informed by a kind of feminism that finds the functioning of the groups "masculine" and "competitive." Having worked with several of these groups during my career, I have to admit that I don't see the groups that I know reflected in this analysis. Are people in the groups competitive? Sure, but they are also cooperative. Are they aggressive? Sure, but they are also mutually supportive. In short, these groups are pretty much like any other group of human beings involved in a complex collective task. If you go in and look for a specific type of behavior you will probably find it, no matter what type it is. If you stop there, without looking for other kinds of behaviors, you're going to come away with a distorted picture of what's going on. In fact, when I read these studies I often wonder if this is how an atom would feel if it could read a physics textbook! So in the end, scientists pretty much tend to ignore postmodern critiques, both because they seem to us to be concentrating on peripheral issues and because when they try to describe things with which we are familiar, they tend to get it wrong.

There is actually another aspect to the Science Wars that I hesitate to bring up, but which also plays a role in scientists' attitude toward the whole postmodern enterprise. Not to put too fine a point on it, there is a feeling that the standards of scholarship in the field aren't very high.

Alan Sokal, a physicist at New York University, became so concerned about this aspect of postmodernism that he ran a little experiment. He put together a long paper titled "Transgressing the Boundaries:

Towards a Transformative Hermeneutics of Quantum Gravity." The paper was written in the supercilious, politically driven postmodern style and was basically a parody. It was hilarious. In fact, I've never known a physicist who could read more than a paragraph of the paper without bursting into laughter—even the footnotes are funny. In any case, Sokal sent the paper off to a journal of postmodern thought called *Social Text*, where it was accepted for publication. After the article had appeared in print, Sokal announced, in an article titled "A Physicist Experiments with Cultural Studies" in a rival journal called *Lingua Franca*, that the whole thing was a hoax. Apparently the journal had been so happy to get something from a physicist who had taken the trouble to learn their language that they published it without asking whether what he was saying made any sense.

What struck me about the aftermath of this whole adventure was the reaction of the *Social Text* editors and their supporters. Instead of engaging in some introspection and asking some basic questions—asking, for example, why they hadn't shown the paper to a physicist—they took Sokal to task for "abusing their trust." In the end, this episode just confirmed the opinion in the scientific community that the whole postmodern critique was not to be taken seriously.

And this is too bad, because in their zeal to prove that science is just as arbitrary and subjective as literary criticism, postmodern scholars have actually widened the Two Cultures gap when it wasn't necessary to do so. Historically, thoughtful scientists have paid attention to what philosophers have said about our discipline. Even today, you often hear references to people like Karl Popper and Teilhard de Jardin when scientists get into wide-ranging conversations. The current gulf between the sciences and the humanities really doesn't do much good for either field and closes off a source of growth for both.

It will probably come as no surprise to the reader that I lay the blame for this modern division squarely at the feet of the postmodern humanists. One of the basic tenets of postmodern literary criticism is that "there is nothing outside the text," which I take to mean that they feel that the proper way to analyze any claim is to look at what is written about it, particularly with a view toward uncovering the author's hidden biases and prejudices. To scientists, who think that the way to test statements is to do experiments to see if they're true, this obsession with text seems irrelevant at best, weird at worst.

To us, the actual words used to advance a claim, like those phone calls to editors described above, have almost nothing to do with its validity, and the hidden preconceptions even less. What matters is the claim itself. Even if it is put forward by a racist, homophobic, misogynistic, dead, White male (have I left anything out?), its truth or falsehood depends on how well it stands up against nature. Once uttered, in other words, a scientific claim is completely decoupled from the person who uttered it, and stands or falls on its own. This fact explains one of the aspects of scientific behavior that nonscientists often find puzzling. Two scientists can get into a roof-raising argument about a particular point in a theory or experiment, then go out and have dinner together as the best of friends. The reason this can happen is that the argument is about a claim that has little to do with the people advancing it, so the argument doesn't have to get personal. (Of course, since scientists are human like everyone else, it does sometimes happen that these sorts of confrontations develop into something personal. My point is simply that they don't have to, and often don't.)

The postmodern fixation on words also explains something else that scientists find disturbing. When I look at what postmodern authors write about subjects like relativity and quantum mechanics (subjects that I routinely teach in the physics department of my home institution), I often find the discussion strange. The authors seem to know some of the buzzwords involved in these fields, but don't seem to have a firm grasp of the subject. When I check the references, I find that the scientific works cited are popularizations rather than the primary literature.

Now don't get me wrong. I have nothing against science popularization—how could I, when I've done so much of it myself? In addition, I would argue that such popularizations have an important role to play in promoting scientific literacy. It is, however, a long way from providing scientific literacy to the average citizen to putting together a scholarly argument intended for an audience of academics. We have a right to demand higher standards in the latter than in the former.

In fact, it seems to me that the very least you can expect from someone who is going to talk about the profound philosophical implications of quantum mechanics is that he or she understand what quantum mechanics is. Like Sokal's hoax, the insistence of postmodern philosophers that they are qualified to talk about difficult subjects on the basis of reading popularizations makes scientists reluctant to take them seriously. So once again we have a Two Cultures gap. This time, however, it is the scientists who are, justifiably enough, tuning out what at least some humanists are saying.

CLOSING THE GAP WITH SCIENTIFIC LITERACY

This postmodern gap, I think, is less serious than the one that existed in Snow's time. For one thing, it took a shorter time to develop and involves only a relatively small number of scholars on both sides. For another, as I mentioned earlier, I sense that the whole postmodern turn is on the way out in many disciplines. But it seems pretty clear to me that Snow was onto something important, something that still has resonances and implications for education today.

Let me start with the education of scientists and engineers. As I mentioned above, there are certainly aspects of the current system I would like to change. As far as closing the culture gap goes, however, I don't think we're doing so badly on this side of the equation. I am not aware of any systematic research on the levels of cultural literacy among scientists analogous to the studies of scientific literacy I discuss in Chapter 6. Nevertheless, when I look at the educational system, I see future scientists getting pretty much the same training in the humanities and social scientists as everyone else. A physics major, for example, will be exposed to as much literature as a sociologist, a chemist to as much sociology as a literature major. My guess, then, is that if and when cultural literacy studies are done, scientists will do as well as other college graduates.

Having said this, I hasten to add that I don't think *any* American college graduates have had enough exposure to either the humanities or social sciences—horror stories like the one about the student who thought that Toronto was the capital of Italy certainly have some basis in fact. It's just that this problem is pretty well spread over all fields of study. There is, in other words, nothing special about the education in the humanities and social sciences that has to be changed only for future scientists and engineers and not for everyone else.

But, to repeat a question I am raising constantly in this book, what about the vast majority? What about the education in science for non-scientists? I would argue that this population is not well served by our current system. Students are allowed to graduate with a fragmentary knowledge of science, then thrown into a society where they have to make decisions about issues involving areas to which they were never exposed. How can you expect someone to comment intelligently on the latest results from the Hubble Space Telescope or the status of string theory if their only exposure to science was a course like Rocks for Jocks (otherwise known as Geology 101)?

To my mind, then, the best way to close the Two Cultures gap is to make sure that all students up to and including those at universities have acquired the basic matrix of knowledge that we are calling scientific literacy. Only then, I think, will science take its proper place next to the rest of the intellectual disciplines as a true part of our culture.

CHAPTER 5

Scientific Literacy: The Argument from Aesthetics

For reasons that have never been completely clear to me, there seems to be a historical assumption that there has to be a conflict between science and the arts. If you think about it for a while, this is a little hard to understand. After all, both disciplines are full of highly creative people engaged in the almost impossible task of trying to make sense of the universe. You would think that there would be at least a kind of mutual sympathy between the two. This is not to say that there are no differences between them—there are. As many scholars have argued, science is essentially a public operation that relies, at least in the presentation of results, on logical reasoning, while the arts are both more private and internal. Perhaps this is what inspired French biologist Claude Bernard's wonderful epigram *"l'art, c'est moi, la science, c'est nous"* (Art is me, science is us).

Certainly, the stereotypes of the scientists and the artist follow these lines—the former buttoned-up, lab-coated, and logical, the latter sloppy, paint-stained, and temperamental. The mere fact that stereotypes like these have limited validity in real life doesn't change the fact that they exist. The question is how to deal with them in terms of scientific literacy.

In this chapter I want to advance two propositions that fly in the face of both the stereotypes and the supposed historical conflict: (1) that "soft" concepts of beauty have always played an important role in the sciences, and (2) that an understanding of science can actually improve the average person's aesthetic appreciation of the beauty of both nature and the works of human artists, and even in some cases, be of assistance to the artists themselves. In other words, not only is aesthetics an important part of science, but the type of knowledge we are calling scientific literacy can also be an important part of the aesthetic experience.

The first proposition, that the concept of beauty—aesthetic criteria—are not foreign to science, may seem a little strange to many people. They are often surprised to learn that you can often hear scientists talk about beauty as an integral part of their discipline. This surprise arises, I will argue, from the fact that the use of aesthetic criteria is so deeply embedded in the sciences that it almost never occurs to us to talk about it to the public.

The second proposition has been the subject of some historical debate, particularly during the Romantic movement of the nineteenth century. Some poets claimed, for example, that a knowledge of science can actually destroy our ability to appreciate the world's aesthetic features. To be honest, I have never met an artist/musician/dancer who felt this way, but the poems are out there, and they get me so irritated that I can't resist saying something about them. I will argue that, far from being a hindrance, science can be a real aid in the aesthetic appreciation of nature, and further that the kind of scientific knowledge best calculated to improve our aesthetic appreciation of the world is precisely the same kind of knowledge we need to function as active and aware citizens—the type of knowledge we have called scientific literacy.

Finally, I will close the discussion with some comments on the direct assistance that science and technology can offer to the practicing artist.

BEAUTY IN SCIENCE

Beauty is in the eye of the beholder, and sometimes it takes rather specialized training to see it. As a lifetime aficionado of football, both as a player and spectator, it's relatively easy for me to pick out the details—the beauty, if you will—in the clash of linemen that just looks like a jumble to my wife, a newly minted (and somewhat reluctant) fan. On the other hand, when I watch an unfamiliar sport in the Olympics—curling, for example—I have no appreciation of the beauty that others tell me is there. I suspect that seeing beauty in the working of science may be the same sort of thing, something that takes some background knowledge to appreciate.

But the fact of the matter is that scientists do, indeed, see beauty in the things they study. Some of this is obvious; the luster of a crystal and the majesty of a distant galaxy are wonderful to behold. I want, however, to talk about a deeper kind of beauty, one not always readily apparent to the senses. I want to talk about the intellectual beauty of scientific ideas.

Darwin closed *The Origin of Species* with the comment that "there is grandeur in this [evolutionary] view of life," a phrase that Steven Jay Gould borrowed for the title of the columns ("This View of Life") that he wrote for many years in *Natural History*. Darwin was arguing that the slow, stately unfolding of evolution over millions of years was, in a very real sense, more beautiful, grander, more compelling than the simple creation story in the book of Genesis. This sentiment had been put forward in 1684 by Thomas Burnett when he wrote in his *Sacred Theory of the Earth:*

> We think him a better Artist that makes a clock that strikes regularly at every hour from the springs and wheels he puts in the work, than he that hath so made his clock that he must put his finger in every hour to make it strike. [1]

Thus after devoting 14 chapters to amassing evidence in favor of his thesis, Darwin closed his argument with an appeal to the aesthetic aspect of evolution. Although his statement is often quoted, the fact that it shows that a scientific theory can be judged by its beauty, at least in part, is often overlooked. As I said above, this omission comes from the fact that the notion is so ingrained in the scientific worldview that scientists hardly notice it, while it seems so bizarre to nonscientists that they just never consider it to be a possibility.

But we don't have to go all the way back to Darwin to see the beauty criterion in operation. Let me take as an example one of the deepest— and, to physicists, one of the most beautiful—theories around. When general relativity was proposed by Albert Einstein in 1916, it marked a fundamental change in the way human beings thought about the universe. (It wasn't an accident that Einstein was named Person of the Century by *Time* magazine in its December 31, 1999, issue.) The deep principle on which the theory is based is as beautiful as it is profound: The only possible universe, Einstein argued, is one in which the point of view (or, to be technical, the frame of reference) of the observer does not affect the laws of nature that the observer sees. In other words, the only possible universe is one in which all observers, regardless of where they are and how they are moving, see the same laws operating.

This is a beautiful and powerful concept, because it allows us to rule out all sorts of possible universes in which we might live. If someone produces a theory, for example, in which a scientist on Alpha Centauri finds different laws operating from those seen on Earth, the principle of relativity allows us to throw that theory out ab initio. It turns out that the concept also implies that the simple mechanical universe of Newton cannot explain a world in which objects move at speeds close to the speed of light. Thus the principle of relativity demanded a major rethinking of the scientific worldview—if, of course, the principle is correct.

And there's the rub, because if you go back to Chapter 1 and review the steps in the scientific method, you see that before we can accept any idea about the world, it has to be tested by observation and experiment. Like every other scientific idea, relativity had to be made to confront nature. If it had failed these tests, it would have become, in words attributed both to the British biologist J.B.S. Haldane and to Thomas Huxley, "another beautiful theory destroyed by an ugly fact." In order to understand what happened after Einstein published his theory, I have to take a short diversion at this point to talk about how a theory like this can be tested. Regard the next few paragraphs, if you will, as an attempt on my part to fill in a small segment of your matrix of scientific literacy.

The problem is that if you look at the technology available in 1916 and then look at the difference between what Einstein and Newton would predict for the outcome of a particular measurement, you find that the predicted differences are much smaller than the accuracy of the available instruments. It would be as if you wanted to differentiate between two theories whose predictions about how long a certain process would take differed by a matter of seconds, but all you had to measure time was a sundial. Obviously, you wouldn't be able to see any differences between the two predictions, even though the differences were there. Except for the few tests listed below, the situation was the same as far as the old-fashioned Newtonian world and the relativistic universe of Einstein were concerned—there simply wasn't enough of a difference between the two to allow them to be differentiated.

In 1916 there were two places where the differences between the two theories were big enough to be seen by available instruments. One involved a small perturbation in the orbit of the planet Mercury. Astronomers had known about the perturbation for decades, but had been unable to explain its source. It turned out that relativity supplied just enough of a nudge to the planet to explain it. Strictly speaking, this sequence of events isn't a prediction—in fact, scientists use the word *retrodiction* to describe a situation in which a previously unsolved problem is resolved by a new theory. Clearly though, the explanation of Mercury's orbit fits into the spirit of the scientific method.

The other difference between Newton and Einstein led, however, to a true prediction. General relativity predicts that when light passes near a massive object like the sun, its path will be bent, so that the source will appear to be slightly shifted compared to that same source viewed when the sun is nowhere around. Normally, of course, we can't see light from a star when it passes near the sun; it just gets drowned out. During an eclipse, however, the stars become visible during the day. In 1919 the British astronomer Arthur Eddington led an expedition to photograph the sky during an eclipse on the island of Principe off the west coast of Africa. He found precisely the kind of shift Einstein had predicted, a fact that, when announced, catapulted Einstein to the status of an international superstar. (Actually, I am continually amazed to discover that he retains this status today. Posters of Einstein are routinely sold in college bookstores, right next to posters of Humphrey Bogart and Marilyn Monroe.)

Incidentally, in keeping with the subject of this chapter, Eddington announced his discovery to some friends with a parody of the *Rubaiyat* of Omar Khayyam:

O leave the Wise our measures to collate
One thing at least is certain, light has weight. . . .[2]

So by 1919 we had one successful prediction, one successful retrodiction, and that was pretty much it. Normally this wouldn't be nearly enough evidence to drive the acceptance of a theory, but in the case of relativity it was. For 40 years the theory was accepted by the scientific community despite the paucity of the experimental confirmation. The reason was simple: Because the theory was perceived to be beautiful, there was a predisposition to accept it. In a sense, the theory's beauty made scientists set the bar for acceptance lower than it would normally be.

You can contrast this situation to one that occurred in the 1950s, when plate tectonics, our current theory of the structure and dynamics of the Earth, came on the scene. The theory wasn't particularly beautiful, and in addition, it contradicted a number of traditional beliefs about the planet. In this case, it took mountains of data to drive the community to acceptance. As it happened, the data were forthcoming and the new ideas triumphed in a matter of a few years. I would argue that the case of plate tectonics is much more typical of the way science works than the case of relativity. In the former that bar was high and a lot of data was required, in the latter the bar was lower because of the beauty of the theory.

Just to make sure that there are no false impressions here, however, I have to add that in recent times many more tests of general relativity have been made, and the theory has come through with flying colors. In the 1950s, for example, technology had progressed to the point where we could detect the (predicted) slight loss of energy in a beam of light as it climbed upward in the Earth's gravity. Today the operation of the Global Positioning System routinely uses the equations of relativity to correct the orbiting clocks that are at the heart of the system. In fact, you could say that in the last half century relativity has passed from the world of theory to the world of engineering.

Of course, the fact that the theory is beautiful doesn't mean that no one has tried to find things wrong with it. During my career I often thought that the whole saga of relativity was like one of those old western movies—you know, the one where a string of young kids come into town to prove that they're faster than the grizzled old gunslinger. In just the same way, one young theorist after another has (metaphorically speaking) strapped on his six-shooter and taken on Einstein, only to have his upstart theory shot to pieces.

Exactly why do we find something like relativity beautiful? I would argue that it's for the same reasons that we find a snowflake beautiful. A snowflake's beauty resides in its symmetry, in that fact that if you rotate it through 60 degrees it looks the same as when it started. For some reason (probably something to do with the deep architecture of the human brain) we find symmetric systems pleasing and beautiful. The principle of relativity embodies a kind of symmetry, too. The universe will look the same, it says, no matter if we view it from Earth, from a spaceship, or from a distant galaxy. Like the snowflake, Einstein's universe is symmetrical.

And just to carry this idea to the present, current theories of the ultimate nature of matter—they go by names like unified field theories, theories of everything, and string theory—posit an even more symmetric universe. They are built on the idea that the only possible universe is one where how we define things like electric charge and other quantities doesn't matter. In effect, they posit a symmetry that requires nature to be indifferent not only to the observer's frame of reference, but to the observer's frame of mind as well. Unfortunately, the mathematical complexity of string theories has thus far prevented their practitioners from making any predictions that can be tested. Because of this, they remain in a sort of scientific limbo. This emphasizes the fact that beauty, in and of itself, isn't enough to produce acceptance of a scientific theory. Scientists need at least a few data points to hang their hats on—as the paleontologist said, data matters!

THE ROMANTIC REBELLION—ART VERSUS SCIENCE

The Romantic movement of the nineteenth century is usually considered to be a reaction to the Enlightenment. As my colleague Richard Florida has pointed out, whenever there is a major force acting in society, there is sure to be a countervailing force as well. Try to make the world more rational and understandable and you're sure to have people come along to emphasize the irrational and mythic. As someone who has often been accused of being a Romantic (it's not always a compliment, especially for a physicist), I have no particular problem with this fact; as I pointed out in Chapter 1, not all questions can be approached by the scientific method. There is, however, one rather minor aspect of the Romantic worldview that I find truly irritating, and I want to deal with it here. This is the notion that somehow, by learning how a particular aspect of nature works, we destroy our ability to appreciate its beauty.

Let me cite two well-known poems that illustrate this claim. The first is from Wordsworth:

Sweet is the lore which Nature brings;
Our meddling intellect
Mis-shapes the beauteous forms of things—
We murder to dissect.[3]

The other is from Walt Whitman:

When I heard the learn'd astronomer
When the proofs, the figures, were ranged in columns before me;
When I was shown the charts and diagrams, to add, divide, and measure them;
When I, sitting, heard the astronomer, where he lectured with much applause
in the lecture room—
How soon, unaccountable, I became tired and sick;
Till rising and gliding out, I wander'd off by myself,
In the mystical moist night-air, and from time to time,
Look'd up in perfect silence at the stars. [4]

To be fair to Worsdworth, scholars point out that the verse quoted above came from a poem addressed to a friend who was something of a bookworm, so that we can be charitable and put it in the "wake up and smell the roses" category of friendly advice. After all, nobody who composed a sonnet titled "Steamboats, Viaducts, and Railways" as Wordsworth did can be considered to be a complete Luddite.

We also have to remember that the nineteenth century was the time when raw industrial capitalism reigned, long before child labor laws or environmental consciousness. Today we can view a railroad bridge in a rural landscape as picturesque, but it's not hard to understand how people then might have felt differently and blamed science for the despoiling of the landscape. And then, of course, there were the factories themselves:

And was Jerusalem builded here
Among these dark Satanic mills? [5]

Whenever I hear these lines of Blake's, I think of hearing my grandfather talk about Chicago at the turn of the last century, when he and his friends would walk six miles to the stockyards each day to see if they had a job. (Oddly enough, he neither shared nor encouraged my own preadolescent rage at the unfairness of that situation; his memories were of the friendship and camaraderie.)

So both my knowledge of history and my own family legends make it easy for me to understand the turning away from rationality and science suggested in these poems. It's easy, in the hustle and bustle of modern life, to turn our eyes back to some imagined golden age when life was simple and pure and blame science and technology for destroying that idyllic world. Indeed, a good deal of modern environmental writing is based on this kind of Garden of Eden theme. Of course, you can quibble about the reality of these visions of the past: When I want to make this point, I often trot out calculations of the amount of manure that would have had to have been produced daily by the horses in New York or London in the "Good Old Days"— it's a big number! But the mere fact that the Garden of Eden may never have actually existed doesn't really matter all that much. It's the dreams that matter, and dreams are the preserve of poets.

It has always been easy to attack the Romantics for their other-worldiness. Perhaps the most devastating satire came from Gilbert and Sullivan in their operetta *Patience*, when the character Bunthorne sings:

> Of course you will poo-pooh whatever's fresh and new
> And declare it crude and mean
> For art stopped short in the cultivated court of the Empress Josephine
> And ev'ryone will say
> As you walk your mystic way
> If that's not good enough for him which is good enough for me
> What a very, very cultivated kind of youth this kind of youth must be. [6]

But having made these easy points, what are we to make of the deeper argument contained in the poems, that understanding something, particularly if that understanding is scientific, in some way ruins the aesthetic experience?

SCIENCE AND THE AESTHETIC EXPERIENCE

One way to approach this question is by citing personal experience. I have spent a good part of my life living with nature: ranching in the Montana Rockies, building a home with my own hands in the Blue Ridge Mountains, and so on. In fact, I would claim a closer association with nature than that enjoyed by most Romantics (and certainly more than that enjoyed by most modern environmentalists). I think I appreciate a sunset or the turning of the leaves in the fall as much as anyone. The fact that I understand why the sky turns red or the chemistry that produces the brilliant fall colors doesn't affect my appreciation at all. In these sorts of situations, the knowledge is simply neutral.

There have been, however, many situations where my scientific knowledge added appreciably to an experience, simply because it made me look at things that might have passed unnoticed without it. I can remember, for example, riding the El in Chicago one winter evening at sunset when I noticed a couple of *sun dogs,* or spots of bright light in the sky to the right and left of the sun itself. They are the result of sunlight refracting from ice crystals high in the atmosphere and, when they are present, are about 22 degrees away from the sun. They follow the sun like a pair of dogs as it sets, which explains the name.

In any case, because I knew what the spots of light were, I was able to call them to the attention of my (nonscientific) companions, thereby making sure that they were able to get a little more out of the experience of the sunset than they would otherwise have gotten. Once they had some basic understanding, their minds could spin up from the dirt and cement of the city to the place where outer space begins, where the stars can be seen during the day, and where the cascades of ice crystals are toying with light from the sun. Their vision of the world was changed, made more meaningful, because they had a little more understanding of what they were seeing.

Actually, my experience has been that using my scientific knowledge to bring previously unnoticed aspects of nature to people's attention is something they really appreciate. I remember one Montana evening when a (very rare) triple rainbow appeared over the mountains during a dance rehearsal. I stopped the rehearsal for a few minutes and herded everyone outside to see it, explaining what it was and why it was so rare. Years later members of that group would recall the experience and tell me how much they had appreciated my comments.

At about the same time, I wrote a series of short columns for a magazine whose primary audience was people who owned and operated trucks—the big rigs you see out on the interstates. I forget the name of the magazine, but the column was about looking at the local landscape as you drove through it so that you could understand a little of its history and how it got to be the way it is now. A trucker going through Nashville, for example, could look at the layers of limestone exposed by the roadside excavations and marvel at the fact that there, a thousand miles from the nearest modern ocean, were rocks made from the skeletons of creatures who had lived in a long-vanished sea. Again, I got enough letters from the drivers to know that they were finding their experiences deepened and changed by the knowledge they had acquired. I even got letters asking about details of local geology that, frankly, I knew nothing about (although I was usually able to find a colleague who could help out).

Even an ordinary experience like an airplane flight can be enlightening with a little background knowledge. For example, on a trip starting from the east coast, the first major geological structure you'll see on your way west is the Appalachian Mountains. If you look at the mountains closely, the first thing that will strike you is that all of the ridges are parallel to each other, with the valleys trending from southwest to northeast. (You can actually see the same thing on an ordinary highway map since major roads follow the valleys.) To understand these ridges and valleys you have to cast your mind back over 300 million years to a time when what is now the continent of Africa slammed into the coast of North America somewhere near the present state of Georgia. The thick layers of rock that make up the Earth's crust buckled and folded up like a wrinkled tablecloth, creating the pattern you can see from the airplane window. When they first formed, those green, rounded hills were as tall and craggy as the Himalayas, but the slow process of weathering has worn them down to their current form. You can trace the "folded tablecloth" as far as the hills of eastern Ohio. As the plane progresses westward, the story of the land beneath you changes, but the fascination remains. Bottom line—you don't have to draw the shade and watch the movie!

All of these examples point to one use of science in the aesthetic experience: The background knowledge draws our attention to things that the untutored eye tends to pass over. What might have been a smudge of light in the urban sky becomes a sun dog. We look more closely and see the faint fringes of color produced when the ice crystals split the incoming sunlight into its constituent colors, much as Isaac Newton did with his prism so many years ago. A nondescript landscape suddenly becomes ordered and understandable, and as a result we start to notice details: the sunlight glinting off the curve of a river, puffy white clouds lining up along the tops of the ridges, or early morning mist still covering the valleys.

It's important to realize that the enhancement of our experiences does not depend in any way on having a detailed scientific understanding of the phenomena involved in the object of our attention. You don't, for example, have to be able to calculate the path of a light beam through an ice crystal to appreciate a sun dog, or know the detailed geological history of North America to enjoy the limestones of Nashville or the Appalachian ridges. The same sort of generalized matrix of knowledge we have defined as scientific literacy is all that is needed to deepen our appreciation of a natural phenomenon and draw our attention to previously unnoticed details.

NEW VISIONS

If bringing phenomena to our attention were the only contribution of science to the aesthetic experience, it would be enough to justify what I am calling the argument from aesthetics. I think, however, that there is another important contribution science can make to the arts—not so much for the general public, but for artists themselves.

Stop for a moment and think about the way that we humans perceive the world. We have five senses, but, with apologies to gourmet chefs and perfumers, the two that matter most for the aesthetic experience are sight and hearing. Both of these depend on the operation of highly developed but very specialized sense organs—our eyes and ears. These organs developed over millions of years of evolutionary history to have one function and one function only: to allow our ancestors to survive long enough on the African savannah to pass on their genes to succeeding generations. Just as our brains did not develop in order to allow us to develop quantum mechanics, our eyes and ears did not evolve to allow us to appreciate an opera or a symphony. All of these organs evolved to allow our ancestors to survive in a brutal world.

Because of this fact, both sense organs have a rather limited range. Consider the eye, for example. From the point of view of a physicist, the human eye is an instrument capable of detecting electromagnetic radiation whose wavelength is between 4,000 and 8,000 atoms long—what we call visible light. It was useful to our ancestors to acquire this sort of detector because (1) we live in an atmosphere that is transparent to these particular wavelengths, and (2) our planet circles a star that puts out most of its energy in the form of visible light.

Because seeing seems so natural and normal to us, it takes a little effort to realize how little our eyes actually tell us about the universe, and how their limitations are molded by our environment. Take the first item listed above—the transparency of the atmosphere. When I think about this aspect of the Earth, I call to mind a cold desert night many years ago when, standing on the mesa at Los Alamos, I could see the street lights of Albuquerque, almost 90 miles away. You have probably had the same experience—think of being in an airplane at night and seeing the lights of a distant city. The fact that we can have this kind of experience means that light is able to travel 100 miles through the atmosphere without being absorbed, an amazing thing when you think about it. Other kinds of electromagnetic waves—microwaves and infrared radiation, for example—can't do this. The atmosphere actually absorbs them rather quickly. Thus creatures who evolved in a different kind of atmosphere might have radio antennae rather than eyes, and would "see" a very different world. Similarly, creatures developing on a planet circling a dim red star might need much larger eyes to capture enough of the longer wavelength radiation to see their surroundings. (This, incidentally, is why pilots of flying saucers are always drawn with this sort of eyes. These large-eyed creatures are called LGM among the cognoscenti—Little Green Men.)

The same can be said about what we hear. Our ears are adapted to pick up sound waves traveling through the air at sea level (or close to it). Had we evolved in a different atmosphere, our ears would be different. The Vulcans on the old *Star Trek* episodes, for example, were given large pointy ears because the air on their imaginary planet was supposedly thin and they needed a larger receiver to capture sound.

So the accident of evolution has equipped us with sense organs capable of detecting only a tiny fraction of what is actually in our world. Similarly, because we are creatures of intermediate size—bigger than microbes, smaller than stars—those same sense organs evolved to detect only other intermediate-sized things, a fact that further limits our view of the universe.

And this is where science can help the artist, because through science we can open up all those worlds and provide new images that are not directly available to our senses. Take the microscope as an example. When Anton van Leeuwenhoek first saw his tiny "animacules" in a drop of pond water in the seventeenth century, he was, in fact, extending the ability of humans to see the world in modes not accessible to the eyes alone. In a similar vein, we are now living in a golden age of astronomy because, for the first time in history, we are able to put observatories into space, above the Earth's atmosphere.

Since the beginning of time, the universe has been sending all sorts of electromagnetic radiation toward the Earth, from radio waves to X-rays to gamma rays. Each of these different kinds of radiation tells us about a different part of the universe in which we live. When this radiation reaches our planet, having traveled (perhaps) billions of light-years to get here, it is absorbed in the last few miles of its flight by the thin layer of air that surrounds us. What a tragedy! But now, thanks to advances in space technology, a tiny fraction of that radiation is captured in orbiting detectors and analyzed, giving us for the first time an almost complete view of the universe. (I'm hedging this statement a little because we can, or will soon be able to, detect other kinds of emissions from distant objects—things like neutrinos and gravitational waves.)

Look at it another way: When Whitman went out to look "in perfect silence at the stars," he was probably able to see a couple of thousand points of light in the sky. This was in the days before so many electric lights lit up the city skies, what astronomers call *light pollution*. It's an impressive experience, one that anyone who is willing to get far away from a city can still enjoy. Nevertheless, as impressive as the night sky appears to the naked eye, it's only a fraction of what's really out there.

Let's start with the solar system. To Whitman, the planets would have looked like bright points of light, with perhaps a tinge of reddish coloring for Mars. How much more he could have seen with even a modest telescope—an instrument his "learn'd astronomer" would have been able to make available. With that instrument he could have seen the rings of Saturn, arguably the most spectacular sight in the solar system. He could have seen the cold, blue-green world of Neptune, the clouds of Venus, the moons of Jupiter. What an expansion of his world that would have been!

And today, with the Hubble Space Telescope in orbit and space probes circling Saturn and on their way to Pluto, he could have seen much more than that. Valleys and basins on Mars bear testimony to the water that once flowed on that planet's surface and perhaps to the life that once existed there before the cold of space crept in. The cracked ice of Europa covers hidden oceans a mile down, and on Titan waves of liquid methane crash into the shores of impossible seas. Turn the telescopes out past our cozy little home system and we find filmy chains of galaxies flung halfway across the universe, with exploding stars and newborn galaxies sending out imaginable bursts of energy. Who would trade this magnificence for the poverty of viewing with the naked eye?

I have to admit that I have sometimes wondered whether painters and other visual artists ever worry about whether they're running out of things to serve as the models for their art. I mean, how many different ways can you paint the front face of a cathedral? By opening new worlds, scientists are, in effect, providing artists with new subjects to interpret. This fact was brought home to me recently when, cruising the Web, I came across a painting titled *Shiva as Telomerase Inside of a Telomere Loop* by artist Julie Newdoll. [7] Clearly inspired by images from modern molecular biology, the painting shows the traditional Hindu goddess Shiva inside a loop representing a *telomere*, the structure that appears at the end of chromosomes. This fusion of ancient lore with the latest electron microscope images seemed to me to be a portent of the future, a future in which the images uncovered by science are incorporated into the visual arts.

Although I have been casting the discussion thus far in terms of artists rather than the general public, there are obvious advantages in our expanded vision of the universe for the latter group. It has been argued, for example, that the image of Earth taken from space, showing our planet as a pale blue dot set in the vastness of space, played a major role in triggering the environmental movement of the 1970s. My own experience has been that cosmic mysteries like dark matter and black holes are just plain exciting for students—they find these sorts of concepts compelling. This is probably why popular books on cosmology sell so much better than books on molecular biology, even though the latter deals with a much more immediate subject. And once again, what one needs to enjoy the benefits of this new vision of the universe is not a detailed understanding of the science involved. You don't have to be able to calculate the rotation curve of a galaxy to appreciate the philosophical implications of dark matter; what you need instead is the kind of generalized matrix of knowledge we have called scientific literacy.

For all of this, I don't believe that the contributions of science to the arts are limited to providing new images. New technologies, new materials, and new ways of producing sound have always been incorporated into artistic techniques. Scholars have argued, for example, that it was the development of new pigments that could last a long time on palettes that made the entire impressionist movement possible. And where, for example, would modern sculpture be without the arc welder and oxyacetylene cutting torch?

The strange ways that new technologies can influence the arts was brought home to me recently when I attended the performance of a new work by my friend Jerzy Sapieyevski. Jerzy is a professor of music at American University as well as a talented composer and pianist. He has a longtime interest in finding ways to unify music with the other arts. In the particular performance I attended, two painters were working on canvases while infrared detectors measured the movement of their brushes. The signals from these detectors were fed into synthesizers to produce sound while the pianist/conductor regulated the synthesizers and wove the musical themes together. In essence, the painters had become both visual artists and musicians—in effect, part of the orchestra.

Now I have no way of knowing whether the paths being explored by my friend and people like him will lead to important new musical movements or whether they will join works like "Symphony for Orchestra, Vacuum Cleaners, and Floor Polishers" in the museum of ideas that failed to catch on. What I do know, however, is that explorations like Jerzy's are the lifeblood of art. If no one is willing to try new ideas, art (like science) stagnates. I also know that without a lot of technology, this kind of experimentation simply couldn't be done.

A FINAL NOTE

Having said all of this, I have to remind the reader (and myself) that scientific literacy can never, in and of itself, provide all of the background needed to appreciate an aesthetic experience. It can be, at best, a bit player in a much larger drama. Nevertheless, considering the larger picture of the aesthetic experience can buttress an important point we have been making about cultural literacy and, by extension, scientific literacy as well.

There is, in fact, a long history of philosophical argument about the nature and meaning of the aesthetic experience. Ideas of category and symmetry abound in classical writings, and we have already discussed the intuitionism of the Romantics. A modern neurophysiologist would approach the description of the experience in an entirely different way, marshalling data on neurotransmitters and patterns in functional magnetic resonance images of the human brain. For my purposes these kinds of abstractions, valuable though they may be in other venues, aren't much use. They shed little light on the kind of education best calculated to enhance one's appreciation of art or nature. Consequently, I decided to take the bull by the horns, as it were, and ask some professional musicians and artists a simple question: What are the characteristics of the person whom you would most like to have in your audience?

As luck would have it, my wife and I were spending a weekend with some old friends in Connecticut. Martin Piecuch is a symphony conductor who has traveled and conducted extensively in Russia, and his wife, Elizabeth Falk, is a director and producer of plays and operas—in fact, the couple was married as part of a performance that included their production of the opera *Gianni Schicchi* and Puccini's *Messe di Gloria*. Because the dinner party included several musicians, an artist, and a young man who operated a thriving business manufacturing harpsichords, I figured this was as good a place as any to get the information I wanted. I sprang my question over a postdinner bottle of wine.

After some spirited discussion, the consensus view was summed up by Martin: "I would like people in the audience," he said, "who have been to concerts before." This simple statement actually summarized a rather complex discussion. What the artists at the table were saying was that their ideal audience would have some familiarity with music, some sense of the background of the pieces they would hear, some expectation of what composers for a particular period would sound like, and some sense of the social and historical context of the music. They didn't expect that the audience would be familiar with the particular music being performed that night, and they certainly didn't expect them to be accomplished performers themselves.

What this group of artists wanted, in other words, was an audience that possessed precisely the kind of generalized matrix of knowledge that I have called cultural literacy. They did not expect their audience to be able to produce music—only listen to it intelligently. In just the same way, I would suggest, scientists should not try to produce a population that can actually do science. What we should aim for instead is an audience that can discuss issues relating to science intelligently. An audience, in other words, that is scientifically literate.

CHAPTER 6

The State of Scientific Literacy

Once we've decided that producing a scientifically literate population is a desirable goal, we are immediately led to ask about the current state of scientific literacy, since this defines our starting point. We must, in fact, start by asking how much citizens know about science right now. This is a difficult question to answer and must be addressed using the skills of the social scientist.

To understand the difficulty involved in gauging scientific literacy, consider this: We know that it is relatively easy to find out what students know about a subject when they finish a course of study—this is, after all, the function of a final exam. But this information, while interesting, doesn't tell us much about what those same students will retain when they read a newspaper a few months (or a few years) later.

Every teacher knows that there is an attrition of knowledge over time. Think back to when you were in grade school, for example. At the end of the summer, how much of the skill in long division that you had acquired so painfully the previous year were you able to summon up on the first day of your new class? At the university level, most science instructors recognize the fact of attrition by assuming that students will have forgotten whatever they learned in their high school courses by the time they matriculate.

In the end, it is knowledge retained, not knowledge acquired for a test, that constitutes scientific literacy, and knowledge retained can't be tested using the techniques of schools and universities. New assessment methods had to be developed to deal with this issue, methods that involve surveying adults whose formal education has ended.

For all of the discussion of science education in this country, it is surprising that the task of finding a quantitative measure for adult scientific literacy has been taken up only recently. The first survey of American adults I could find was conducted in 1957 under the auspices of the National Association of Science Writers. By modern standards, the design was simple and not very well thought out. Participants were asked to talk about four issues that were current at the time: fluoridation of water (who can remember that?), strontium 90 (a product of atmospheric nuclear weapons tests), polio vaccine (then newly introduced), and space satellites (a rarity at the time). The problem with this kind of survey, of course, is that today's burning issues quickly become yesterday's headlines. No one would bring up issues like fluoridation and strontium 90 today, and satellites and vaccines are so commonplace that they scarcely enter the modern consciousness. From the point of view of measuring scientific literacy, the problem with this sort of test is that there is no way of comparing a 1957 response to a question about fluoridation to a modern response to a question on stem cell research. Since establishing a baseline against which change can be measured is one of the prime tasks of the social scientist, this sort of survey just doesn't get us very far.

JON MILLER AND THE BIRTH OF QUANTITATIVE SCIENTIFIC LITERACY ASSESSMENT

The application of social science methods to the problem of assessing scientific literacy is a field that is generally reckoned to have been founded by Jon Miller, now John Hannah Professor of Integrative Studies at Michigan State University. Jon is a big man, with a fringe of white hair and an imposing brow. He would be intimidating if it weren't for his sense of humor, his ready laugh, and his seemingly endless supply of good stories. We have collaborated on various scientific literacy projects in the past, so when I talked to him about including his work in this book, we arranged to get together the next time he came to Washington on business. That is how, on a rainy autumn night with Jon still slightly jet-lagged from his recent trip to China, we sat down at my favorite Lebanese restaurant to discuss how one goes about founding a new field of intellectual inquiry.

"I didn't start out being interested in scientific literacy," he said.

"In fact, my Ph.D. thesis (at Northwestern) was a study on the way that children acquire the notion of nationality." Jon actually began his graduate work at the University of Chicago, where he took his master's degree and learned how to conduct surveys. Along the way, he also acquired the standard mathematical skills of the modern social scientist. As a result, when he decided to take some time out from his studies to get some real-life experience in government, he was snapped up immediately by directors in what are now the Office of Management and Budget and the Office of Science and Technology Policy. "Later on, they told me that they needed a guy who could count," he laughed, "but it wasn't like [the TV show] *West Wing*—they only allowed us on the first floor of the White House."

He delights in telling a story about a particular event from this period in his life. He had to present the proposed budget for programs he was supervising at a formal hearing, and his request included a number of new posts for food inspectors at ports. He knew that the director of the Budget Office, who was conducting the hearing, was an inveterate coffee drinker and was seldom without a cup in his hand. Consequently, when the item including the inspectors came up, he launched into an explanation of how coffee beans were dried in Africa, by simply being laid out in the sun for a period of time. He pointed out that in this situation, birds could be expected to fly overhead and would, as birds do, add a certain level of contamination to the beans. The extra inspectors, he asserted, were needed to make sure that none of these beans got into the American market. Staring suspiciously at his coffee cup, the director growled to his assistant: "See that the inspectors stay in."

Jon enjoyed his life in the government, but one of his supervisors gave him some very good advice: "If you don't leave and get a Ph.D., you'll always be an assistant." So it was back to grad school.

It was at Northwestern that he first became interested in scientific literacy. He was dating a woman with far-left political views ("I worked for the Kennedy campaign, she worked for Eldridge Cleaver") while at the same time playing tennis with a group of guys who were studying to be nuclear engineers. At the time, the issue of nuclear power was very much in the forefront of public attention, and his friends got into intense arguments about it. ("They wound up throwing food at each other.") Intrigued, he read up on the subject. "I discovered that you actually had to know some science to deal with the issue," he recalled.

"If you had to vote on something connected to nuclear power and didn't know the science, the ballots might as well have been in Urdu." This insight, in turn, led him to ask the kinds of questions about what one needs to know to function as a citizen that I discussed in Chapter 3. Being a social scientist, he also began to ask how people acquire that knowledge—a learning process that Jon thinks of as a piece of what is called *political socialization.*

After a stint on the faculty at Chicago State University, he moved to Northern Illinois University in DeKalb, a town in the far penumbra of Chicago, as associate dean of the graduate school. "At this time, there was money (from the Carter administration) for projects that used unemployed people to do 'good things' for local governments," he recalled. He used this money to hire people to carry out manpower surveys in several neighboring counties, as well as surveys of high school and college students on the acquisition of citizenship. He hired many part-time workers ("It was like running the world's largest McDonald's"), so he was always on the lookout for new projects to keep them busy. He soon realized that he had created a general purpose survey laboratory—which he named the Public Opinion Laboratory—and eventually negotiated a transition from his deanship duties to the full-time direction of his new survey unit.

By the late 1970s, he was coming to the realization that public debate in America was going to become increasingly dominated by scientific issues and that he now had an organization capable of carrying out surveys of scientific literacy. He got a grant from the National Science Foundation to work on the project. "We began in an attic, with old carrels that the library was throwing away. We actually used pencil-and-paper forms for the first survey." Over the next few years he worked out the basic form of the questionnaire and started collecting data on what was to become the *Science and Engineering Indicators Program*—a program that has now been running for 20 years and constitutes the gold standard of scientific literacy time lines.

By this time he had moved his operation into a refurbished hotel building next to the university and was thinking about expanding the operation to include international collaborations. The first of these was a 1988 cross-cultural survey between the United States and Great Britain. "Once we had the U.S.-U.K. data" he said, "everyone else fell into line. We now have data from more than 40 different countries, all using the essentially the same core set of questions." In fact, it was this international aspect of his work that explained his jet lag that night in Washington.

You don't run into a lot of surprises when you compare surveys in America and European countries, but when I asked Jon about some of the problems he was having in China, the stories got interesting. "Their data are heavily weighted toward men," he said. "Husbands won't allow their wives to talk to the surveyors, and the Chinese scientists needed help dealing with this problem." Actually, this sort of gender problem crops up in other countries, and many survey teams now include women to interview their counterparts "unbeknownst to the husbands."

In addition, in China there is a large population of people (about 100 million by some estimates) who are registered in rural areas but actually live in the cities. Since this is illegal, such people are often reluctant to divulge honest demographic information to strangers.

But in spite of these sorts of problems, the work of putting together a worldwide assembly of data on scientific literacy goes on every day—a long way from a grad student food fight and old carrels in an attic.

CONSTRUCTING THE TEST

Jon thinks of scientific literacy (he actually uses the term *civic scientific literacy*) as composed of three elements: (1) a knowledge of basic scientific constructs, (2) an understanding of the processes of science, and (3) an understanding of the impact of science on society, although he uses only the first two in his cross-cultural comparisons. Because he was concerned with establishing a long timeline of responses, he chose to test for concepts whose importance would not change over time, concepts related to what I call Great Ideas in Chapter 12.

Questions on the survey can be either open-ended or require a short answer. For example, one of the open-ended questions on scientific constructs begins this way:

When you read the term DNA in a newspaper or magazine, do you have a clear understanding of what it means, a general sense of what it means, or little understanding of what it means?

If the person indicates that he or she has a "clear understanding" or "general sense" of what the term means, the questioner then continues with this question:

Please tell me, in your own words, what is DNA?

The responses, recorded verbatim, are then coded using standard double-blind procedures, procedures that have been shown to produce highly reliable results. As you might expect, these sorts of open-ended questions give the best measure of understanding of particular topics in the scientific literacy field.

Because interviewees can terminate the interview at will, particularly if the interview is conducted over the telephone, the number of open-ended questions that can be asked is limited. Consequently they are supplemented by short-answer questions. For example, the following might be asked in a true-false format:

Lasers work by focusing sound waves.
All radioactivity is manmade.
The earliest humans lived at the same time as dinosaurs.
The center of the Earth is very hot.

Some might be simple fact questions:

Which moves faster, light or sound?

Testing for understanding of the scientific process is somewhat more difficult, since the subject doesn't lend itself to true-false or multiple-choice formats. One approach Jon used was to follow the same format as in the DNA question above, asking whether the interviewee understood the process of scientific inquiry and then, for positive responses, asking for an explanation. (The most common response to this question was to say that science involves experimentation, which Jon feels indicates an acceptable minimal understanding of the subject.) Specific questions aimed at probing the understanding of probability, the difference between astrology and astronomy, and the role of control groups in experiments are included in various versions of the test.

A typical American survey will involve about 2,000 participants. In the European Union, the scientific literacy questions are usually embedded in a larger semiannual survey called the Eurobarometer, and involve about 1,000 participants for each country in the survey.

People who score above 67% on the tests of scientific constructs and display a minimal understanding of the scientific method are labeled as *scientifically literate*, or *well-informed*, while those who did well in one category but not the other are labeled *partially scientifically literate* or *moderately well-informed*.

As I mentioned above, one use of a test like this is to keep track of national trends in scientific literacy over time. Thus it becomes a kind of scoreboard that tells us how we are doing in our efforts at improvement. Ideally, a test designed to do this is sufficiently well thought out so that it can be given in essentially the same form over a long period of time, thus making comparisons over time easy to interpret. At the same time, including common questions in tests given in different countries allows us to make cross-national comparisons as well.

RESULTS OF THE SURVEYS

Looking at Jon's data over the last 20 years, two important results stand out:

1. Average American scientific literacy scores have been increasing over the past decade.
2. Americans tend to be at or near the top in international comparisons.

Given the poor showing of American high school students on international achievement tests in science and math, and given the lack of improvement on standardized tests in elementary, middle, and high school, these results are little short of astonishing. Furthermore, since Jon's tests are administered to adults and not to students, we might expect some attrition in the level of understanding during the adult years, an expectation that makes the relatively good performance of the adults even more puzzling.

Let me give you some numbers to illustrate the two points listed above. In the early 1990s, the number of American adults whom Jon classified as *scientifically literate* was about 12% of the population. In his 2005 survey, that number climbed to 28%, with tests in intermediate years showing a steady improvement.[1]

There is no cause for complacency in these results; that 28% of Americans meet Jon's (admittedly minimal) requirements for scientific literacy means that roughly three Americans in four do not. This result does not inspire confidence in our ability to conduct sophisticated public debates in the years ahead. On the other hand, the trends show that we are doing something right, and make it possible for us to raise the question (as I do in Chapter 12) of how to improve on this base.

The international comparisons are equally surprising. In the same 1990s surveys that showed American scientific literacy at 12%, only Britain was close (at 10%), followed by Denmark and the Netherlands (at 8%). In the most recent survey, the United States ranked second, trailing only Sweden (at 35%).[2]

Confronted with these long-term trends in the data, one's first response is to ask why they should be as they are. I mean, given the poor showing of American students on achievement tests, how can American adults show improvement in scientific literacy and look so good in international comparisons?

When I put this question to Miller, he had a ready response. "The best indicator of success on these tests is the number of college-level science courses taken by each individual. What we are seeing here is a result of the fact that Americans are required to take science courses at the university, while Europeans and Asians are not." He attributes the rise in scientific literacy scores, in large part, to the increase in courses aimed specifically at nonscientists in American universities—what we will call the Physics for Poets and Integrated Science movements in Chapter 8.

When these courses are properly designed and presented, they can provide students with the tools to read and understand new scientific developments over the next several decades of their lives. In fact, many college-level science courses for nonscience majors provide invaluable insights into the nature of matter, the development and structure of living organisms, and the nature of our home planet and the universe in which we live. To the extent that these courses are able to help students learn about the nature and structure of DNA, for example, they provide a tool for those students to read and think about genetic modification issues for decades into the future, even though the exact content of future disputes cannot be known in the present.

In fact, Jon's surveys identify two positive effects of college-level science courses. One, not surprisingly, is that students who have been supplied with the basic background of scientific literacy retain some of that knowledge later in life. The other is more indirect but, in the end, almost as important as the direct knowledge imparted by the courses. It turns out that people who have had a science course in college are much more likely to avail themselves of the multitude of informal science education opportunities in our society. They are more likely to read books, magazines, and newspaper articles about science; more likely to visit museums; more likely to watch televisions programs like *NOVA;* and so on. They are, in other words, more likely to engage in lifelong learning once they have the basic foundation that enables them to do so. The growing audience of people who seek out information about science has, in fact, fueled the growth of what we may broadly call *science journalism,* a point to which we will return in a moment.

But whatever the reason, Jon's huge international database not only defines the starting point in our quest for increased scientific literacy, but gives us some indication of how that quest might proceed. It also tells us that the pessimistic tone of writers like Morris Shamos, who in his 1995 book *The Myth of Scientific Literacy* despairs of making any progress at all in this area, is perhaps a bit premature. It turns out that we're not doing that badly.

ASSESSING SPECIFIC SCIENTIFIC LITERACY PROGRAMS

Now that we know the tools at our disposal for measuring scientific literacy in the adult population, we can turn to the problem of evaluating specific programs, certainly a prime concern in our standards-based era. There are two issues that, as far as I can tell, will make this task difficult, if not impossible.

The first of these, alluded to above, involves the attrition of knowledge with time. If we accept Jon's analysis of the reason for America's relative good showing on scientific literacy surveys, then *when* someone studies science must rank with *what* is studied in importance. It is, for example, reasonable to expect that someone who has recently read a *Time* magazine article on genetics (and there have been many lately) will do well on the sample DNA question given above, while someone who read an equivalent article a couple of years ago might not. This means that it isn't just formal course work we would have to know about, but recent informal exposures to science as well. This would be a daunting task for a volunteer 20-minute survey.

And this brings us to the second point. When we measure an adult's level of scientific literacy, we are looking at the integrated effect of a lifetime of learning, both formal and informal. It is virtually impossible to tease out something like the effect of an innovative third-grade science course or a particular museum exhibit from this sort of data. Even if people could remember the sort of instruction they received in elementary school, the effect of a particular course will be confounded by all of the other variables in the problem.

What this means is that in assessing specific programs, we are going to have to fall back on the more traditional final-exam model, even though it doesn't give us the information we really want. We can argue that someone leaving a particular experience with a high level of scientific literacy is likely to retain more knowledge later on, simply because he or she has more to begin with. In any case, this, coupled with Jon's data charting the overall success of our enterprise, is all the data we will have at our disposal.

A TECHNICAL ASIDE: ITEM RESPONSE THEORY, TEST DESIGN, AND INTERCULTURAL COMPARISONS

Every once in a while, it's a good idea to look at data a little more deeply than we have been doing so far in this chapter. This sort of exercise can give the reader some sense of the real-world complexity that hides behind data that appear simple on the surface. In this section, then, I will present something called item response theory (IRT), a theoretical method for analyzing test and survey results by looking at responses to specific questions or "items." The reader who wishes to skip this somewhat technical section can pick up the discussion at the end of the chapter without loss of continuity.

In evaluating the knowledge possessed by people participating in a given survey or test, it is often possible to get a lot more information than that contained in the raw test score. We can, for example, analyze individual responses to see how well the test or questionnaire is designed, thus allowing us to discount responses to poorly designed questions. In addition, it often happens that different versions of the same test are given to different groups; for example, tests given in different countries and over a long period of time are seldom identical to each other. Because of this fact, it is extremely useful to have a way of comparing results despite the differences in the details of the test. IRT is a theoretical method designed to deal with this sort of problem.

An "ideal" question or item on an "ideal" test would work something like this: People who had achieved a certain level of understanding of the subject being tested would get it right, while those who had not achieved that level of understanding would get it wrong. Thus the answer to this question, by itself, would tell us whether the person being tested had arrived at that particular level of knowledge.

The problem with this ideal situation is that it requires us to have some independent way of knowing a person's level of knowledge. This may be possible in some cases (for example, in a test of reading comprehension where there are other tests of reading level available). Usually, however, the state of a person's knowledge is what we are trying to measure, so we don't have it ab initio.

The usual way we get around this problem is to use the person's overall test score as a proxy for the person's state of knowledge. In this case, we can make a graph for our ideal question like the one shown in Figure 6.1. On the horizontal axis we plot the overall test score, while on the vertical axis we find the percentage of test takers who answered the question correctly.

In this graph, we see that no one who scored below a certain number on the test was able to answer the question, while everyone who scored above that number was. We have labeled this critical overall test score, which marks the transition point for this question, the *threshold*.

In general, depending on the question, this threshold can be anything from 0% to 100%. The first of these possibilities would correspond to a question that virtually everyone can answer: "Does hot air tend to rise or sink?" The second would correspond to a question that no one can answer: "Explain the concept of renormalization in quantum field theory." (Actually, if your sample included a theoretical physicist you might get a correct response on the second question, but such an inclusion is statistically unlikely.)

Figure 6.1. Hypothetical Answers to a Question on an Ideal Test.

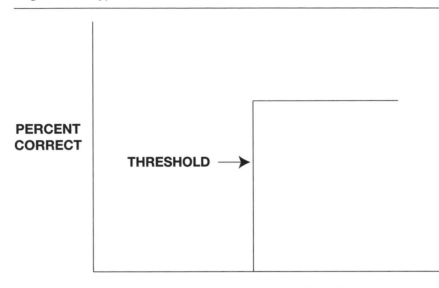

Both the 0% and 100% case are useless from the point of view of analysis, since they do not allow you to make any sorts of discrimination between test takers. If everyone can answer a question, or if no one can, you learn little about the correlation between a specific kind of knowledge and other factors such as education.

On the other hand, when the threshold is between these two extremes, as in Figure 6.1, it differentiates between the test takers: Those who answer it correctly will have total scores above the threshold, while those who fail to answer it will have lower scores. An ideal test would consist of a series of items with progressively higher thresholds, so that almost everyone could answer the first question and almost no one could answer the last. With such a test, it would be relatively easy to deduce the effects of things like age, gender, and educational level on the results.

Of course, no "real" test question or item would ever give results like those shown in Figure 6.1. The actual responses to a specific item will look more like the curve shown in Figure 6.2, which is called the *item response curve*. Instead of the sharp step we talked about for the ideal case, this result shows a gradual rise of correct answers with test

Figure 6.2. Item Response Curve: Typical Answers to Questions on a Real Test.

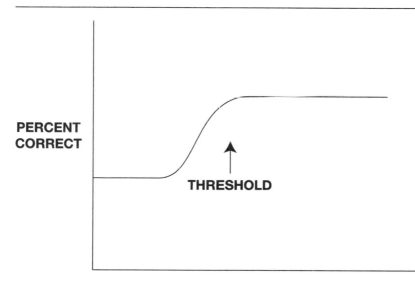

PERCENT CORRECT

THRESHOLD

OVERALL SCORE

scores, a more realistic situation. There are several points about this curve that need discussion.

First, the curve shows that there will be a certain number of correct answers even for people who do poorly on the test, and who therefore (presumably) know nothing about the subject being tested. The reason for this is that on multiple-choice or true-false tests, someone guessing at random will get correct answers a certain percent of the time. This state of affairs is reflected in the name statisticians give to the place where the curve crosses the vertical axis—they call it the *guessing parameter*. On a multiple-choice test with four possible answers, for example, the simplest estimate of the guessing parameter would be 0.25—a chimpanzee taking the test would probably get one answer in four correct. (Actually, the parameter is a little more complicated than that because the more someone knows, the more he or she can eliminate incorrect answers and reduce the number of possible guesses.)

The second point to note is that, while this curve does show an increase in the percentage of test takers who provide a correct answer with increasing knowledge, it doesn't have the sharp threshold seen in the ideal case of Figure 6.1. We can deal with this by noting two characteristics of the curve—the spot where 50% of the test takers answer the question correctly, again called the threshold, and the steepness of the curve at the point. Obviously, the steeper the curve is, the more closely it approximates the ideal. These three numbers—the guessing parameter, the threshold, and the slope—are taken to characterize the curve.

Why would you need to analyze test results using IRT? Jon Miller gives an interesting example. When a question on probability and genetics was included in a test given in Japan, the subject matter was changed from one involving the probability of birth defects (as in the American version) to one involving the probability of color in flowers (which, for some reason, was more consistent with with Japanese educational practices). How could you tell whether these two different forms of the question were comparable in difficulty?

"It's like judging diving in the Olympics," Jon said. "Each dive is given a difficulty level which is used to weight the judges' scores of the dive itself. This means that a diver who got 10s from the judges on a simple dive may actually wind up below a diver who got lower scores on a more difficult dive. In the same way, IRT gives us a way of comparing the difficulty of questions on tests."

Like most mathematical techniques, IRT has many uses. To bring one example close to home, Jon uses the SAT as an example. "This test is given on 16 consecutive Saturdays," he explains, "so that if you were motivated to do so, you could send people into the early sessions to memorize questions. By the end, you could actually have a copy of the entire thing. If you want to avoid this possibility, you actually have to give 16 different tests." And this, of course, means that you have to know, item by item, that the tests are scored so that students taking a slightly more difficult version aren't penalized. In a world where a few points on the score can have an effect on college admissions (not to mention lawsuits), having a reliable way of doing this is essential. It is precisely this sort of comparison method that is provided by IRT.

The point I would like the reader to take away from this discussion of IRT is this: Behind the simple overall test scores that we read about in the newspapers, there are a variety of techniques, of which IRT is one, that allow for a much more sophisticated analysis of the results. In the end, there is a lot more to analyzing answers to questionnaires than just toting up right and wrong answers.

CHAPTER 7

The Research Pipeline

Over the years I have stumbled over a number of aphorisms to which I have given the tongue-in-cheek name of Trefil's Laws. The one that is appropriate to the subject of this chapter goes like this:

Whenever someone figures out something about how the universe works, sooner or later someone else will come along and figure out how to make money from that knowledge.

SCIENCE AND TECHNOLOGY

As a scientist, of course, I wish that people would appreciate my craft for its beauty, for the depth of understanding it adds to their lives, for its sheer intellectual brilliance. I've been around long enough, however, to know that this just isn't the way things are. Whenever people in leadership positions, particularly those in government or business, talk about the value of science, it doesn't take long for the subject to slide over into the economic benefits that go along with scientific advances. If I had a nickel for every time I've heard a governor talk about building the next Silicon Valley in his or her state, I'd be a rich man. (Okay—that may be an exaggeration. At the very least, though, I'd be able to get another cup of the excellent cappuccino I'm sipping while I write this.)

So when we ask the question, "Why science?" we have to face the fact that the way ordinary citizens, as well as political and business leaders, think about the question is quite different from the way it is approached by scientists and educators. To the former, science is valuable because it leads to economic gains and an increase in human health and happiness. To the latter, science is, or at least should be, part of a liberal education—something that is to be studied for its own sake, rather than for any economic gains that it might generate.

This situation reminds me of a conference I attended years ago. It was held in Richmond, and was devoted to the state of liberal education in the Virginia university system. The conference involved all of the universities in the state, and I was there as one of the representatives of the University of Virginia. For 2 days speaker after speaker excoriated what they saw as the rampant professionalism of that generation of students, the decline of the liberal arts, and the lamentable (in their view) ascendancy of science and engineering in the academy. Then, as a grand keynote, the governor came over to address the assembly. From his first sentence, all he talked about was how universities could serve as nuclei for the creation of new technologies and new business, citing the Route 128 corridor near Boston and (yes) Silicon Valley as examples of the kinds of things he wanted Virginia universities to produce. I have never seen a clearer example of the broad gulf that separates academics from the rest of the population.

I have to admit that, as one of the token scientists at the meeting, I felt a certain sense of wicked glee while I listened to the governor's talk, particularly after having bitten my tongue for the last day and a half. But the more I thought about it, the more I began to realize that the two different points of view expressed at that conference really weren't so different after all. Science is both a generator of economic activity and a part of a liberal education, and always has been. In fact, the point of the Trefil's Law above is that when you start applying the scientific method to problems in nature, even if economic benefits are the farthest thing from your mind, your work will inevitably generate those benefits whether you want it to or not.

There are so many historical examples that make this point that I scarcely know where to begin. Perhaps the early part of the nineteenth century is as good a place to start as any. This was a time when the scientific method was being applied to an old but rather mysterious set of natural phenomena—those involving electricity and magnetism. Although these phenomena had been known from ancient times, their systematic study had really only started in the late 1700s, and by the 1830s it had already become apparent that, although electricity and magnetism appeared to be quite different from each other, in point of fact they were intimately connected. The British physicist Michael Faraday was exploiting these connections in a series of stunning experiments that produced the prototypes of both the modern electric motor and the electrical generator.

Faraday's life story is particularly interesting. He was the son of a blacksmith, a man who was a member of a small Christian sect. In early nineteenth-century England only members of the Church of England could attend universities, so formal educational paths were closed off for him. The practice of the day was to apprentice young men so that they could learn a trade; in this case, Faraday's father apprenticed him to a bookbinder. Faraday read the books as he was binding them and discovered a deep interest in science. A client gave him some tickets to a series of public lectures given by Humphrey Davy, a prominent chemist. Faraday took notes on the lectures in a beautiful Victorian copperplate hand, bound the notes in leather, and presented them to Davy when he applied for a job as his assistant. He was hired, and eventually worked his way up to a position of prominence in English scientific circles. He even became a frequent guest of Queen Victoria at court.

The story I want to recount concerns a day when Faraday was giving the British prime minister a tour of his laboratory. (Since I have heard this story told about almost every British prime minister from 1815 on, I will leave the question of which one it really was to the historians.) After being shown Faraday's new devices, the prime minister is supposed to have said. "Well, Mr. Faraday, this is all very interesting, but what good is it?" Faraday answered, "What good is it? Why, Mr. Prime Minister, someday you'll be able to tax it!"

The point of the story is that neither the prime minister, with all of his political acumen, nor Faraday, arguably the greatest scientist of his generation, could possibly have guessed where those clumsy pieces of apparatus would lead. From Faraday's point of view, he was just trying to figure out how nature worked, and the devices were more or less accidental and unanticipated outcomes of that search. Neither man could have foreseen the modern world, where the energy stored in coal or uranium or falling water is used to spin massive generators, producing current that is sent out over entire continents through a web of power lines.

Nonetheless, it's hard to imagine modern society without the benefit of Faraday's devices. Take something as simple as getting into your car, pressing a button, and having the window come down. It is a small electric motor, a distant descendent of what the prime minister saw that day in Faraday's lab, that moved the window. You've probably used electric motors a hundred times today, without even knowing it. Any device with a fan—computer, hair dryer, air conditioner—uses them. In fact, the great electric power network that girdles the globe is a splendid, though often unappreciated, result of Faraday's work. This statement has particular meaning to me as I write this, because a week ago a series of freak storms ripped through the Washington, D.C., area, toppling trees and cutting off power to hundreds of thousands of homes. After living through 2 summer days without electricity, my neighbors and I came away with a renewed appreciation of Michael Faraday and his contributions to human welfare.

In retrospect, it's easy to see the importance of Faraday's work, and we would quickly realize the utility of modern research designed to advance our ability to generate electricity—think about work on the development of high efficiency solar panels, for example. In 1830, however, as the prime minister's remarks show, it was far from obvious that research into the arcane areas of electricity and magnetism would ever do anything other than satisfy the curiosity of the researcher. There is no question in my mind, for example, that if you had asked Faraday if his work would lead to a better way of lighting houses than oil lamps, he would have had trouble understanding what you were talking about. Yet within the next 24 hours, everyone reading this book will flip a switch and turn on an electric light, thanks to the genius of that same Mr. Faraday.

So, in some way, scientists' curiosity and the utility of their work seem to be linked. And that brings us to the question of how, exactly, these two different aspects of the scientific enterprise are related, and how one flows inevitably into the other through what I call the *research pipeline*. In the standard language of science policy, this question has traditionally been posed in terms of three different kinds of scientific work: *basic research, applied research,* and *development*. I should point out that up to this point in the discussion I have been somewhat loose in using the terms *science* and *technology*. The two, although related, are not identical, and it is in following the research pipeline that the difference between the two becomes clear. In essence, basic research clearly represents science, while development represents technology. It is in the middle of the pipeline, in the area I'm calling applied research, that a transition occurs between the two.

BASIC RESEARCH: KNOWLEDGE FOR ITS OWN SAKE

If you're talking about scientists based in universities—the sort of scientists who seem to be in the news most often—then the chances are that they would say that they are doing something called *basic research*. The standard way of defining this activity is to say that basic research is research undertaken solely to learn more about nature, to push a particular boundary of knowledge farther into the unknown. But while this definition has the advantage of being simple, it focuses more on the state of mind of the researcher than on the research itself.

How Science Grows

When I think about basic research, I prefer to think about the way that science grows and expands over time. I envision this process as being something like the growth of a tree. At the heart are those grand principles that have been verified over and over again, and which no one seriously questions. The conservation of energy, the basic laws of electricity and magnetism, and evolution by natural selection all fall into this category. They are the core of our understanding of our world, the basic skeleton on which the rest of science hangs. As I argued in Chapter 1, these ideas are so solidly established that we can take them, for all intents and purposes, as given.

This does not mean that the core ideas of science are impervious to change. It's just that for them change takes the form of placing limits on their validity, rather than replacing them with something new. Take the effect of the advent of relativity and quantum mechanics in the early twentieth century on the perceived validity of Newton's laws of motion as an example of this process. In 1905 a young patent clerk in Bern published a very strange paper. Highly philosophical, with almost no reference to data, it proposed that a grand principle operated in the universe. Called the principle of relativity, it holds that the laws of nature are the same for every observer in the universe, regardless of the observer's state of motion. With this paper, Albert Einstein started a profound revolution in the way that scientists look at the universe.

But it's important to realize that Einstein's work in no way replaced Newton's. It simply delineated the region where Newton's laws were valid. Since the experimental backing of Newton's laws involved slowly moving objects, those laws could still be applied to those objects. What Einstein showed was that you can't take laws that apply in one area—slowly moving objects—and automatically apply them to another—objects moving near the speed of light. When I want to make this point to my students, I talk about the experience of getting on an airplane at an American airport. When you get on the plane, everyone around you is speaking English. You cannot assume, however, that the same rule will hold when you get off the plane— English might or might not be spoken where you land. In the same way, we cannot assume that Newton's laws, derived for slowly moving objects, will apply when they are extended to regions other than those for which they have been verified.

At the risk of belaboring this point, let me talk about numbers. One of Einstein's most famous predictions (which has, by the way, been massively verified by experiment) is that moving clocks will slow down. Fair enough, but if you use Einstein's equations to calculate how much your clock will slow down if you drive at 60 miles per hour, you quickly discover that you would have to drive for much longer than the lifetime of the universe to see your watch slow down by one second. Only atomic clocks, capable of measuring time to 13 decimal places, can actually discern the differences between Newton's and Einstein's descriptions of the normal world. This is why we can still use Newton's laws to design skyscrapers and send space probes to the outer planets.

In the end, the result of Einstein's work is that we see Newton's laws as still being valid for the range for which they were originally verified, but not outside that range. For the record, an exactly analogous argument can be made about extending Newton's laws to objects the size of atoms. In this case, the new laws go under the name of quantum mechanics, and were developed in the 1920s and 1930s. As was the case with relativity, when you apply the laws of quantum mechanics to normal-sized objects moving at normal speeds, you get virtually the same results as you would from Newton.

At the very core of the scientific tree, then, are the grand principles like Newton's laws and the principle of relativity, and it is the goal of every science to push its own laws and ideas into this charmed circle.

Moving away from the core principles, we enter a kind of gray area where ideas are still being tested—where they are still on probation, as it were. I think of this as the outer wood of the tree. Moving still farther out, you come to an exciting region where growth is happening, where new ideas are being born and dying in a grand intellectual mêlée. I think of this as being like the cambium under the bark of the tree, the place where new wood is actually produced.

By far the greatest part of basic research takes place in this region, where new phenomena pop up every day and everything is up for grabs. It's an exciting place to work, and it's small wonder that scientists compete ferociously for the relatively few available places out there at the edge. This is also the place where virtually all of the news stories about science originate, and this sometimes leads to problems. If the only contact you have with science comes from these news sources—a situation that describes most people in our society—then you are likely to get the impression that scientific knowledge is a very unstable thing. After all, what else can you conclude if you see a result announced in blaring headlines one day, only to be contradicted by other results, heralded with equally blaring headlines, the next?

There was a time when the main task of a teacher was to try to convince his or her students that scientists don't know everything, that there are still vast areas of ignorance out there. Because of the kind of news coverage we have, however, our task has shifted to trying to convince a new generation of students that scientists know *something*. If you think about the analogy between scientific knowledge and a growing tree, the previous generation of students concentrated on the heartwood and was unaware of the cambium, while today's students think that the cambium is all there is.

The Results of Basic Research

It is out there at the fringe of understanding, at the interface between knowledge and ignorance, where basic research is concentrated. And just as Faraday couldn't have foreseen the outcome of his research on the connection between electricity and magnetism, today's researchers are often at a loss to give more than the most general predictions about where their work will lead. Thus one characteristic of basic research is an innate difficulty in predicting what its effects will be.

Take the field of elementary particle physics, where I started my career, as an example. This is a field devoted to the most basic kind of research, encompassing as it does the nature of the ultimate constituents of matter and energy. Research in this field is expensive, requiring huge accelerators whose costs run into the billions of dollars. Today, with the imminent completion of a machine called the Large Hadron Collider (LHC) at the European Center for Nuclear Research (CERN) in Geneva, scientists are hopeful that we will soon have an understanding of the ultimate constituents of which the universe is made. Just as Faraday and his colleagues broke through to an understanding of the nature of electricity and magnetism, modern scientists hope to break through to an understanding of matter.

Where will this lead? There is absolutely no way of knowing. We can, however, speculate a little. We know from Einstein's famous equation $E = mc^2$ that matter and energy are related, so one possible outcome of an understanding of the nature of matter could be new sources of energy to drive our society. Can I guarantee that this will happen? Of course not. All I can do is say that Trefil's Law has always worked in the past, so that a century from now people will undoubtedly be looking at us the way we look at Faraday and his colleagues, marveling at their ignorance of the massive changes they were about to create in the human condition.

Let me give one final personal illustration of the way that people in basic research often underestimate the impact of their work. I had the good fortune to be a grad student at Stanford during the 1960s, when future Nobel laureate Arthur Schawlow was developing a newfangled gizmo called the *laser* in a basement lab in the physics building. A large, jovial man, Schawlow delighted in having brown bag lunches with grad students. I remember one session where he was asked what uses his device would have outside of basic research. He thought for a moment, and then opined that the laser could possibly be used to make an improved tool for correcting mistakes on manuscripts being produced on electric typewriters!

I should interject here that this doesn't mean that we can't make rational decisions about funding and directing basic research—federal agencies and private corporations do that all the time. If you want a particular piece of knowledge (e.g., some properties of a distant galaxy), you can probably figure out a way to get it. The point is that once you have it, you often have no way of guessing the changes that that knowledge might ultimately make in human lives.

So basic research often proceeds in the dark as far as the prediction of ultimate outcomes is concerned. On the other hand, the grand principles that form the core of science are the result of basic research carried out by past generations of scientists. Every one of them began as a lightbulb going off in one scientist's head as another idea out there at the fringe of the unknown. Bringing the idea from the fringe to the core is another task of basic research, somewhat less glamorous, but arguably just as important.

And finally, in keeping with the eternal testing to which scientific ideas must be subjected, there are a few hardy souls engaged in trying to disprove, or at least limit, the grand principles at the core. In general, this tends to be a high-risk–high-reward operation. After all, the reasons these principles are at the core is that they have withstood the testing of generations of scientists. Nevertheless, there is always the prospect that they may be shown to be wrong in some area where they haven't yet been tested. A particular law—Newton's law of gravitation, for example—may have been tested to five decimal places, but maybe there's something new in the sixth. Or maybe it's been tested at the scale of the solar system and the scale of the laboratory, but not at the scale of miles. Again, there's always a chance that someone will find something new. (Actually, both of these are examples of attempts to disprove the law that wound up verifying it, in effect, closing up a gap in our knowledge.) In fact, the only core principles that are still undergoing vigorous testing are those involved with Einstein's theory of general relativity. This is because testing the theory is technically challenging, and it's only in the last few decades that our instrumentation has advanced to the point where some tests can be made.

In the end, then, basic research can be found all across the spectrum of science. It generates new ideas out at the fringe, tests and winnows those ideas over time, and keeps testing even the most hallowed principles. Having said this, I repeat that most basic research is done out at the frontier—that's where the action is, and that's where most basic researchers spend their time. A smaller cadre works on developing and testing probationary ideas, and a hardy few spend their careers looking for holes in accepted ideas. The threads that run through all this work are (1) a lack of detailed knowledge (and often a lack of interest) in where the work will eventually lead, and (2) a desire to push the frontiers of knowledge farther out, regardless of the practical utility of the results. The question, then, becomes this: How does this sort of research lead to the kind of world-changing results we've been talking about?

APPLIED RESEARCH: GETTING READY

Laws of nature run the gamut from grand, overarching principles shared by all the sciences to detailed results that may apply only to a single material or a single condition. The laws by themselves, however, can seldom be applied directly to a specific human need. There has to be some spadework done to take something from the laboratory to the point where it can be put to use, and that task is the job of applied research.

Although I've been involved in a few applied research projects during my career, I decided to get a real professional perspective from Jeff Newmeyer, a man who has had a long and distinguished career managing research at a major aerospace company. In addition to being a lifelong friend, Jeff was also my first Ph.D. student at the University of Virginia.

"There are basically two situations," Jeff said. "There are situations where the technology is in place and people are looking for problems it can solve, and then there are situations where there is a well-defined problem looking for a technological solution." The real skill in managing research, according to Jeff, consists in putting two disparate groups of people together. On the one hand, there are the scientists and engineers who are concerned with developing new technologies; and on the other hand there are the marketing and mission people who are concerned with making a particular project work. Not surprisingly, bringing the grand principles of science to bear on real problems involves as much in the way of management and people skills as it does technological expertise.

When I asked Jeff to give me an example of how trying to solve a practical problem would lead scientists to deal with the fundamental laws of nature, he thought for a moment, then said, "Miniaturization." If you think about modern electronic devices, you realize that they have been getting smaller and smaller for a long time. When I first began using computers in grad school, for example, they were huge machines that took up several rooms, and it required a small group of acolytes to keep them operating. Yet those machines, now consigned to museums, had much less computing power than the 3-pound laptop on which I am writing these words.

Or consider cell phones. The original phones in the 1980s were the size of bricks—part of the fun of watching reruns of old TV shows is seeing people using those huge, ungainly devices. They too are entering the realm of nostalgia, to be replaced by phones that not only fit in the palm of your hand, but can take pictures and may, for all I know, calculate your income tax. The rapid progress of miniaturization was brought home to me recently when I got an (unsolicited) e-mail invitation to an 1980s party at which anyone who actually brought one of the old phones would get a free drink!

Let me take a slight technical diversion at this point to look at the scientific problems presented by miniaturization. The area involved is *quantum mechanics,* which is the study of the behavior of particle motion at the atomic and subatomic levels. As I mentioned earlier, this science was developed in the 1920s and 1930s as an exercise in basic research. By 1947, however, three physicists—Walter Brattain, John Bardeen, and William Shockley—had built the first transistor. This is the basic working tool of modern computers and, indeed, all modern electronics. The three shared the 1956 Nobel Prize for this invention. In fact, Bardeen is the only person to be awarded two Nobel Prizes in physics (the other was in 1972 for the development of the theory of superconductors).

That first transistor was an ungainly thing, the size of a golf ball. But as the years went on, transistors got smaller and smaller as scientists and engineers got better and better at arranging the atoms from which they are made. In fact, the progress has followed something called Moore's law, named after Gordon Moore, one of the founders of Intel. This law (which is really more of an observation) says that every characteristic of computers—size, computing power, memory, and so on—will become twice as good as it is today within 18 months. This law seems to have worked for the size of electronic equipment, to the point that today your computer or cell phone may have hundreds of thousands of transistors on a chip the size of a postage stamp.

But if you follow Moore's law to its logical conclusion, you realize that before too long, the size of transistors will shrink to something comparable to the size of atoms (this is actually expected to happen sometime around 2020). This is clearly going to be a crisis of some sort, a crisis that is already foreshadowed in work on miniaturizing electronic devices. What does it mean, for example, to have an electrical current flowing in a device of atomic dimensions, when the particles in the current are of the same dimensions as the wire through which they are flowing?

To understand this problem from the point of view of the scientist, think about an ordinary piece of material, such as the paper in this book. In the paper virtually all the atoms find themselves surrounded by other atoms. Only a tiny percentage is actually at the edge of the paper. The behavior of an atom surrounded by other atoms has been pretty well worked out and is part of the normal paraphernalia of quantum mechanics. As components in electronic devices get smaller and smaller, however, the percentage of atoms at the edge begins to grow; indeed, as we approach the limit of Moore's law, all of the atoms will be near an edge, and the standard results of quantum mechanics will cease to apply.

This means that scientists will have to start exploring a new regime, a regime involving systems with small numbers of atoms. And while some of this work clearly falls under the heading of basic research, a lot of it does not. It's important to understand that this sort of research isn't really the kind of thing that led to the laws of quantum mechanics in the first place. Those laws apply to systems with a small number of atoms as well as those with a large number, and were derived by scientists looking for the laws that govern the behavior of all matter. The problem is to take those laws (which are, remember, the product of basic research) and apply them in an area that no one has previously thought about—the area of systems with only a few atoms.

As with any human activity, it is sometimes hard to draw a sharp boundary between applied and basic research. However, in this case, if the research involves looking at specific materials and devices in an industrial laboratory, and if the researchers are interested in solving a specific problem rather than discovering general laws, we would be pretty safe in categorizing the work as applied. It is somewhere in this process, I think, that we can talk about switching over from science to technology.

Of course, once the applied scientists and engineers have solved the technical problems standing between them and their goal, their results still have to be turned into a useful device. This final step in the research chain is called development.

DEVELOPMENT: THE PAYOFF

The first thing to realize about the development phase of any project is that it is here, for the first time, that nonscientific criteria begin to play a dominant role. Scientists and engineers may develop a wonderful alloy for making a car body, but if using that alloy doubles the price of a car it's unlikely to find widespread use. In this case, the nonscientific criterion of cost comes into play. There are other such criteria: efficiency, size, weight (think aerospace), and durability. The job of development (or, as it is sometimes called, R&D, where *R* stands for "research") is to see that the final application of the scientific principles results is a product that does the job people expect it to.

To get a feel for how this kind of technology works, I talked to Robert Blonski of Ferro Corporation, a major supplier of specialty chemicals for industry. In addition to being my brother-in-law, Bob has parlayed his Ph.D. in materials science into a distinguished career in development—a career that includes at least 10 patents. A big, serious guy, Bob likes nothing better than to lean back, sip a beer, and talk about pigments.

Pigments?

Yes, pigments. At first, this may seem like a strange topic for a book on science, but take a look around you. How many of the things you see have their natural color? Chances are that almost everything in view has been dyed, painted, or colored in some way, and that means that they have been treated with pigments. Color is very important to us. After all, humans evolved as primates, which means that we are highly visual. This is why we say "I see" when we actually mean "I understand." From the time of the Neanderthals, we have used artificial colorings to decorate ourselves and our surroundings.

But it's a long way from those primitive daubings to modern pigments. It's one thing to use pigments for decoration, but they have evolved for other uses as well. For example, when you paint your house you pick a color to make it look good, but you also expect the paint to protect the house from the weather. You expect the paint to adhere well to the outer walls—you don't want it peeling off in a few weeks. Thus you quickly go from judging a paint by its color to look-ing at other properties—durability and adhesion in this case. Suppose you want your paint to stick to the fuselage of an airplane when it goes through the sound barrier? Or suppose you want the paint to make the airplane invisible to enemy radar? Or to have the kind of deep coloring you see on modern cars? It's clear that you aren't going to get those kinds of paints by grinding up nuts and berries, and that's where the development process comes into play.

When I asked Bob how development was done these days, his first reaction was very interesting. "Things have really changed over the past 30 years," he said. He went on to explain that most large corporations these days don't do their own basic research: "In the old days big companies would have research labs that were just like the ones you'd find in a university. Today, everything starts with market needs. Basically, the research and development departments have been decentralized, and technological people are integrated into product development groups." Indeed, in some companies, these groups include economists whose job is to project future market needs. Bob points out that this system, which seems very foreign to academic scientists, has been enormously successful, and has produced all sorts of what he calls "stuff"—plasma TVs, cell phones, and all of the other paraphernalia of modern life.

But doesn't the emphasis on product development prevent com-panies from planning for the long term?

"There are three levels of planning," said Blonski. "Horizon 1 deals with things we need right now. Horizon 2 deals with things we'll need in the next 2–3 years, and Horizon 3 deals with 3–5 years. Very few corporations make serious plans past seven years. Basically, if a new technology comes along and we decide that we need it, we can either license it or buy the company that owns it outright."

In other words, a division of labor seems to have grown up in the development world. New ideas and new technologies—in effect, the tail end of the applied research process and the being of development—are largely done by small start-up companies, often spun off from university research. Large corporations then monitor these new techniques and acquire them as needed to produce specific products.

This turning away from research by large corporations has long been decried by academic scientists, but the system seems to work pretty well. There is certainly no dearth of good ideas. Blonski pointed out that every Tuesday morning the U.S. Patent Office publishes about 3,500 new patents and, for good measure, publishes more than 6,000 patent applications every Thursday. "I have to get up early in the morning," he joked, "so that I can get online before the site gets jammed up."

In addition, large corporations often have the capital and expertise needed to pursue new ideas. For example, after Ferro had developed a technique for producing fine powder coatings, their scientists realized that those same techniques could be used to produce new types of pharmaceutical products. As a result, the company bought a smaller manufacturer of pharmaceuticals and made a major move into a new area. "The key," said Bob, "is leveraging your technology to follow market needs."

In a world where that depends so much on the development of new technologies for economic growth, the division of labor strategy described above seems to do what it is supposed to do—keep the product pipeline full. And that, of course, is where basic and applied research finally produce the payoff that most people have come to expect from science.

THE LIGHTNING ROD: A CASE HISTORY

One of the great advantages of looking at historical examples is that you know, in retrospect, what all the right answers are—an advantage denied to the actual players in the drama. So in order to illustrate the great research pipeline flowing from basic to applied research and from there to product development, I will turn to one of the earliest examples of its operation, the development of the lightning rod. This example has the added advantage of having all of the crucial steps being carried out by the same individual, our own Benjamin Franklin.

Franklin, like many men of his time, dabbled in many different areas. He is best known for his role in the foundation of the United States, of course, but, like his younger friend Thomas Jefferson, he has so many firsts associated with his name that we today, living in a more complex age, can only marvel at the breadth of his contributions. He founded the first public subscription library and the first fire company, as well as the American Philosophical Society, the first learned society in the New World, all after having built a successful business career as a printer. He was an accomplished inventor and tinkerer, producing both bifocals and the Franklin stove, as well as a plethora of less well publicized inventions including, believe it or not, swim fins.

In the late 1740s Franklin was already a well-established businessman and had the leisure time to turn his attention to science. This was a time when the scientific method was starting to be applied to the phenomenon of electricity (the people involved in this research went by the amusing title of "electricians"), and Franklin's interest was piqued by attending a talk given by an itinerant British lecturer. When some electrical equipment was donated to his library, he set to work to improve it. In those days the only way of producing electrical charge was to rub two materials together, typically glass and some kind of skin or cloth. Franklin built an apparatus in which a large glass cylinder was turned against pieces of buckskin. This apparatus could build up quite a large charge and produce pretty impressive sparks when it discharged.

This can be related to a common experience. Have you ever walked across a thick carpet on a dry day, reached for a doorknob, and had a small spark jump from your hand to the door? This, in miniature, is the kind of thing Franklin and other "electricians" saw in their apparatus. Franklin realized that the spark and its accompanying audible snap was similar to lightning, with the spark playing the part of the lightning bolt and the snap being analogous to thunder.

Up to this point, Franklin was engaged in basic research. What are the properties of electrical discharges, and are they related to lightning? The famous kite experiment, when he and his son flew a kite in a thunderstorm and observed sparks jumping off a key at the end of a cord was an attempt to provide experimental verification for the identification of lightning with electricity.

To understand what happened next, you have to know something about the dangers of lightning in the growing cities of the eighteenth century. At that time, lightning strokes were the source of devastating fires, particularly in America, where the buildings were largely made of wood. There was no understanding of what caused lightning, and they were seen as acts of God in the full sense of the phrase.

This lack of knowledge led to all sorts of superstitions, some of which had fatal consequences. In 1767, for example, the authorities in Venice decided that it was sacrilegious to suppose that God would allow lightning to strike a church, so they put their stores of gun-powder in a church vault. When lightning hit the church steeple (a particularly inviting target) the resulting explosion killed thousands of people and wiped out a whole section of the town.

So pervasive was the belief that churches were somehow immune from lightning that in France it was felt that the best way to avert lightning strokes was to have someone go up into the church steeple and ring the bells. Indeed, some bells were cast with the inscription *Fulgura frango* (I break up the lightning) on the bells themselves. Needless to say, this belief resulted in the deaths of many bell ringers—it's hard to imagine a worse place to be in a lightning storm than a church steeple.

So at this point, Franklin had acquired an understanding of the nature of electricity and was aware of a pressing societal need for protection from lightning. What followed was a quick progress through applied research and development to the production of the lightning rod.

The first question to ask in this situation is simple: If a lightning stroke is, indeed, a flow of electrical charge, then how can that flow be diverted from buildings? This requires some research into materials suitable for carrying large electrical currents. Franklin quickly settled on iron, a cheap and readily available metal. Finding this material would be an example of applied research.

Franklin had now established the basic principle of the lightning rod: The stroke of lightning would be diverted to the ground through a piece of metal. In effect, the rod provides an easy path for the electrical charge that has built up in the thundercloud to get to the ground. He had also settled on iron for his device. The next question (the development phase) involved finding the best arrangement to make the system work. In the end, Franklin settled on a design involving a sharp piece of brass fitted to the top of an iron rod. Here are his words:

> It had pleased God in his Goodness to Mankind, at length to discover to them the Means of securing their Habitations and other Buildings from Mischief by Thunder and Lightning. The Method is this: Provide a small Iron Rod. . . of such a Length, that one End being three or four Feet in the moist Ground, the other may be six or eight Feet above the highest Part of the Building. To the upper End of the Rod fasten about a Foot of Brass Wire, the Size of a common Knitting-needle, sharpened to a fine point. . . . A House thus furnished will not be damaged by Lightning. [1]

Once Franklin made his proposal, there was still some development to be done. For example, the question was raised in England as to whether a pointed wire would work better than a blunt one. (We know now that the pointed wire works best.) There were issues about how many rods were needed to protect large buildings, the best way to arrange lightning rods on ships, and so on.

But more interesting than the technical developments were the ways that different countries reacted to Franklin's new invention. Some, like Italy, adopted it immediately; others, like England, were somewhat slower; and some, like France, resisted it. It is the reasons for the resistance that I find interesting, because I see them prefiguring modern responses to technologies like cloning and genetic engineering.

One set of arguments centered on the question of whether the lightning rods actually made things worse than they would otherwise have been. It was suggested by some English scientists, for example, that the rod actually caused the lightning bolt to form in situations where it would not have formed in the absence of the rod. Some of the pointed versus blunt argument was cast in these terms. (Today, we can say that there is a grain of truth in this argument, in the sense that the lightning bolt is directed toward the rod, but in the absence of the rod would strike somewhere else.)

A deeper argument, advanced by some theologians, was that by diverting the lightning bolt, Franklin was, in effect, thwarting the will of God. In their eyes, Franklin's invention, far from being a boon to mankind, was an act of great impiety. This argument has been advanced over and over again as science has advanced—for example, it was used to argue against immunizations. In an age when the accusation of "playing God" has been hurled at scientists developing biotechnology, it is somewhat comforting to realize that this is an old and venerable human reaction to changes in the way we deal with the world.

DIRECTING THE PIPELINE

Once we understand the process by which scientific ideas get converted into tangible benefits, it is natural to ask whether the research pipeline can be directed to produce certain desired results. How many times, for example, have you heard someone say, "If we can go to the moon, why can't we. . . ?" The answer to this question, it turns out, depends on which stage of the research pipeline you're talking about.

Before getting into that discussion, however, I want to make a historical disclaimer. What I have presented as the research pipeline, the orderly movement from basic to applied to developmental research, is an idealization. Like the scientific method discussed in Chapter 1, it presents a simplification of what is actually a rather complex process. Sometimes the development of a technology showed up first, to be followed later by a deeper understanding—the steam engine is a good example of this sort of inversion. My sense, however, is that today new results are much more likely to flow from basic research—and thus follow the research pipeline as I've defined it—than to come from technology.

If you are at the business end of the pipeline, using the results of basic and applied research to develop a useful product or device, directing research efforts is fairly easy. If, for example, you have a round widget and you want to produce a square widget, chances are that you'll be able to do it by a judicious assignment of money and personnel. Every year millions of products, from airplanes to cosmetics, are created in precisely this way. At this end of the pipeline, in other words, we have a high level of control over what we will produce.

Moving upstream, to the area of applied research, the situation becomes a little more uncertain. We can take the Apollo program, which put the first man on the moon, as an example. When this program was announced by President Kennedy, we already had done pretty much all of the basic research we needed to do to complete the job. There were, however, a huge number of pieces of applied research that needed to be done, from the development of heat resistant materials to shield spacecraft on reentry to the production of food concentrates to keep the astronauts well nourished during their trip. Each of these required its own applied research program, and each, in the end, was successful.

One unanticipated outcome of the Apollo program was the large number of spin-off products that it produced. From concentrated orange juice to the plastic in the wheels that we use in our luggage or rollerblades, the products of the space program are all around us. Science advocates often use this as an example of the payoff from research when they want to argue for increased research funding.

Indeed this argument is actually a source of disagreement between me and Jeff Newmeyer. Over late-night drinks, we often tangle about whether examples of spin-offs should be used to justify research funding. Jeff's point is that if you want to develop concentrated orange juice, you don't build spaceships—you develop the orange juice. My argument is that until we have a product, we don't really know that we have a need. I doubt, for example, if a development program for luggage would have produced the material in roller wheels. This is one of those issues on which reasonable people can, and do, differ. In any case, you now know both sides of the argument and can make up your own mind.

There is no question, however, that when we move from applied to basic research, our ability to predict and control where the research will lead drops dramatically. In fact, this drop-off is so dramatic that I was able to use it as one of the defining characteristics of basic research above. Another example of a crash government program—the War on Cancer—illustrates this point. While the Apollo program shows how successful a massive influx of money can be in carrying through an applied research program, the War on Cancer (initiated by Richard Nixon in 1971) illustrates how inappropriate it is to try to do the same in basic research.

The term *cancer* is used to classify a collection of diseases characterized by uncontrolled cell growth. To understand and control this disease, we will have to understand a great deal more about how cells work than we do now, and certainly more than we knew back in 1971. From the political point of view, the goal of the program was to wipe out the disease. I was involved in experimental cancer research at the time, and no scientist I knew believed that it was possible to do this. We simply didn't know enough about the operation of cells. This was, after all, a full quarter of a century before the Human Genome Project.

Because achieving its goals would have required massive advances in basic knowledge, the War on Cancer never achieved the kind of dramatic results that we saw in the Apollo program. This doesn't mean that it was a failure; over the intervening years we have acquired more and more of the necessary knowledge, and my sense is that the field of cancer therapy now stands poised for dramatic advances over the next decade. The war simply illustrates the difficulty involved in trying to direct the flow of basic research.

In academic circles, this feature of basic research is sometimes brought up as a kind of supercilious criticism of the scientific community. We are supposed, somehow, to see that our research will lead to some evil end (usually the development of weapons), and therefore abandon it. I have several difficulties with this kind of argument. For one thing, I fail to see that improving the national defense by working on weapons is wrong; indeed, I have, during my career, worked in a number of classified research areas. But more important, even if I did think that weapons development was wrong, there is simply no way of telling, before the fact, how a particular piece of basic research will evolve as it moves down the pipeline to final product. To comply with my academic friends' supercharged morality, we would have to stop doing research altogether, which means we would have to give up all the good that could come from it. In the words of my friend Julian Noble, a nuclear physicist at the University of Virginia, "Withdrawal is no more the solution to the research problem than it is a solution to the population problem."

To sum up, then, the possible outcomes of research become more and more knowable as we move down the pipeline from basic research to development. In order to ensure that the pipeline keeps delivering the output that drives our society, we need to keep all parts of it healthy. When people ask me why the public should support some seemingly pointless basic research, I always use a horse racing analogy to reply. Basic research is like a horse that has paid enormous dividends every time we've bet on it in the past. Although this doesn't guarantee that it will do so in the future, you'd be a damn fool to bet against it next time around.

CHAPTER 8

The Historical Struggle with Science Education

There is always a temptation when we are immersed in a contemporary problem to think that we and our colleagues are the first to have encountered it. In fact, my sense is that there is almost no problem, particularly a problem in education, that hasn't reared its head at some time in the past and that our predecessors haven't grappled with in one form or another. This is certainly true of the problems connected with science education. In this chapter I would like to talk about the way that scholars and educators in the past have thought about this issue for the simple reason that many of the new ideas in science education actually have a long pedigree, and knowing about their history often gives us a useful perspective in dealing with them today.

NINETEENTH-CENTURY ENGLAND

Whenever the educational system fails to deliver the knowledge about science that people feel they need, alternate methods of delivery will spring up. In contemporary America, this parallel system is composed of books, magazines, newspapers, broadcast media, and museums. This system flourishes today, I believe, because of the poor design of our science curriculum. In nineteenth-century England, the educational system had a different kind of failure. Few people were actually able to attend school in those days, and those who did were not exposed to much in the way of science. Whatever the cause of the failure, though, the end result was the same then as it is now. The discoveries of scientists were opening vast new intellectual vistas, and people wanted to know about them. The main mode of delivery that sprang up in nineteenth-century England was the public lecture, and many of the famous scientists in Victorian England moonlighted as lecturers. Michael Faraday, the era's leading physicist and the quintessential self-educated man, was particularly famous in this regard. (His biography is sketched in Chapter 7.) The so-called Christmas Lectures he started at the Royal Institution are still offered to large audiences every year. He also used to lecture at factories, and there are stories of thousands of workers pouring out into the yard to hear him. My favorite example of Faraday's genius as a lecturer concerns an outdoor talk on a freezing cold day, when he rubbed two blocks of ice together, melting them and thereby demonstrating that heat is a form of energy and not an intrinsic property of materials.

The main problem faced by scientists in the formal educational system was the death grip that classical scholars had on the curriculum. Herbert Spencer attacks that situation in his 1855 essay, "What Knowledge Is of Most Worth":

> While what we call civilization could never have arisen had it not been for science, science scarcely forms an appreciable element in our so-called civilized training. [1]

Actually, Spencer's essay was a little over the top on this issue. He seems to have been advocating an educational system based almost exclusively on science, rather than taking the much more reasonable (and defensible) view that science ought to have been incorporated into the system along with other studies. Perhaps this overreaching was what caused "nine out of ten" of his countrymen to find his proposals, in his words, "simply monstrous."

A quarter century later, Thomas Huxley echoed Spencer's ideas in a lecture titled "Science and Culture":

> Neither the discipline nor the subject matter of classical education is of such direct value to the student of physical science as to justify the expenditure of valuable time. . . [and] for the purpose of attaining real culture, an exclusively scientific education is at least as effectual as an exclusively literary education. [2]

Again, I think it is best to interpret these statements as an attempt to move the British educational system away from its exclusive attention to classical studies and more toward the model of the research university that was being developed in Germany at the time. I was fortunate enough to have studied at Oxford as a Marshall Scholar many years ago, so I have some memories of the lingering English attitude toward classical studies at that institution. I was plunked down, a wide-eyed kid fresh from a blue-collar neighborhood in Chicago, at Merton College, which was founded in 1272. I can still remember the Warden of the college (who billed himself as the "last living Hegelian philosopher") telling me how fortunate I was that the college had decided to drop the requirement that incoming science students demonstrate proficiency in Latin. (To this day I'm not sure whether he was glad of the change or expressing regret over the decline of standards since the Good Old Days.) He also explained, quite seriously, that the British Empire had been built by men who were trained to write classical Greek poetry and then sent out to run the railway system in Calcutta. His attitude, then, was that it really didn't matter what you taught the best and the brightest—their native intelligence would allow them to excel wherever they went.

What can I say? The system seemed to work pretty well in the nineteenth century. The problem, of course, is that paying attention exclusively to students at the very top of the academic ladder won't produce the vast army of technically trained people needed to run a modern society, and it certainly won't produce a scientifically literate citizenry. In modern England, changes in the old universities and the building of new ones (the so-called red-brick institutions) have been undertaken largely to deal with these problems.

THE AMERICAN EXPERIENCE

In America, science education had a somewhat different history. As was fitting for a new country still in the throes of development, Americans from Jefferson and Franklin on saw science as a tool for producing practical results, as a way of getting to what I called in Chapter 7 the "business end" of the research pipeline. Indeed, Alexis de Tocqueville commented that

> In America the purely practical part of science is admirably understood. . . but hardly anyone. . . devotes himself to the theoretical and abstract portion of human knowledge. [3]

This leaning toward the practical was clearly reflected in the American educational system. The first school devoted to the study of science and technology was Rensselaer Polytechnic Institute in Troy, New York, an institution that enjoyed then, as it enjoys now, a well-deserved reputation in engineering and applied science. It wasn't, in fact, until the founding of Johns Hopkins University in Baltimore in 1876 that America acquired a research university on the European model, one clearly devoted to what we have called basic research. At its opening ceremony the first president, Daniel Coit Gilman, stated the goals of the university this way:

> What are we aiming at? The encouragement of research. . . and the advancement of individual scholars, who by their excellence will advance the sciences they pursue, and the society where they dwell.[4]

Later in his speech, Gilman stated what has become the accepted wisdom—the mantra, if you will—of the research university as far as teaching is concerned:

> The best teachers are usually those who are free, competent and willing to make original researches in the library and the laboratory. The best investigators are usually those who have also the responsibilities of instruction, gaining thus the incitement of colleagues, the encouragement of pupils, the observation of the public.[5]

It is one of those strange historical coincidences that the founding of Johns Hopkins in 1876 came in the same year that America's first great theoretical physicist, J. Willard Gibbs of Yale, began publishing his work—work that the practical bent of American science prevented his colleagues from appreciating until after his death. Gibbs was an interesting character. Tall and aristocratic (his father had been a professor, too), a lifelong bachelor (he lived with his sisters), he was known as a kindly, if somewhat abstract, teacher. According to my old thermodynamics professor, tradition has it that he told only one joke in his entire life. His sister and he were hosting a dinner party, and when she started tossing the salad, he took the tongs and said, "Let me do that, my dear—after all, I'm the expert on inhomogeneous equilibria." I guess you had to be there.

In any case, once science started to get a toehold in the American system, it doesn't seem to have encountered the kind of resistance that

Huxley and his colleagues railed against in England. The restructuring of American medicine that was triggered by the "Hopkins men" was just one example of the rapid application of science to the outside world by pragmatic American scholars following the introduction of the research university.

JOHN DEWEY AND THE SCIENTIFIC HABIT OF MIND

Our concern here, however, is less with the advancement of science in general than with the way that science was introduced into various levels of the curriculum. It should come as no surprise that the first theoretical justification for incorporating science came from the pen of John Dewey, a man who had a profound effect on virtually all aspects of American education.

In the early part of the twentieth century, Dewey formulated a rationale for general science education that still resonates in the educational establishment today. Arguing for the inclusion of science in the high school curriculum, he maintained that one goal of education should be to produce what he called a *scientific habit of mind*. His main concern at the time was the high school curriculum, but it is important to remember that in those days the percentage of Americans attending high school was quite low. In 1910, for example, the percentage of Americans who were awarded high school diplomas was below 10%—a shockingly small number by modern standards. In fact, it wasn't until 1940 that this rate passed 50%. Consequently, when we listen to Dewey discussing the high school curriculum, we have to remember that he was talking about an elite body of students, perhaps equivalent to undergraduates at prestigious, highly selective universities today. Universal high school education, education for everyone rather than a few fortunate students, was well in the future when he wrote.

Dewey's arguments for science education in high school are similar to what I called the argument from civics in Chapter 3.

> Contemporary civilization rests so largely upon applied science that no one can really understand it who does not grasp something of the scientific method. . . . on the other hand, a consideration of scientific resources and achievements from the standpoint of their application to the control of industry, transportation (and) communication, not only increases the future

social efficiency of those instructed, but augments the immediate vital appeal and interest of the subject. [6]

He also argued that learning science was valuable in and of itself because of the mental development that went along with it:

> The formation of scientific habits of mind should be the primary aim of the science teacher in the high school. [7]

During the 1920s and 1930s, it was this argument, rather than Dewey's original argument from social utility, that seems to have captured the attention of educational philosophers, and it was this argument that largely drove the inclusion of science in the curriculum. I have to say that, reading these books several generations later, I am struck by both the idealism and the naiveté of the authors. The idea that there was a magic bullet called the scientific method that would miraculously turn every student into a logical, reasoning human being certainly has not been borne out by experience. This has not, of course, kept this same argument from popping up over and over again in our modern debate—a point to which I will return in some detail later.

You can get some sense of what educators in the 1930s thought they were about by looking at a definition of a successful student given by University of Wisconsin educator I. C. Davis:

> We can say that an individual who has a scientific attitude will (1) show a willingness to change his opinion on the basis of new evidence; (2) will search for the whole truth without prejudice; (3) will have a concept of cause and effect relationships; (4) will make a habit of basing judgment on fact; and (5) will have the ability to distinguish between fact and theory. [8]

Wow! I think we would all be happy if our students acquired even a fraction of Davis's characteristics, much less all of them. Again, experience has shown that these kinds of unrealistic education goals simply aren't going to be met. This doesn't mean that we can't keep them in front of us as worthy ideals, of course. It just means that at some point we need to get our heads out of the clouds and ask ourselves what we can realistically expect from the students who actually sit in our classrooms.

People in the "Wisconsin School" put together a number of tests of "scientific attitude" in the 1930s, but the results didn't seem to correlate with students' abilities in science or with their general scholastic achievement. Thus, by the start of World War II, the first great push to introduce Americans to science through the school curriculum came to a somewhat disappointing end. The one great achievement of this movement—one whose importance we should not underestimate— was that from this time forward, science was locked into the American system of universal education.

SPUTNIK AND ITS EFFECT ON AMERICAN EDUCATION

I don't think I'll ever forget my introduction to the Space Age in 1957. I was an undergraduate at the University of Illinois, walking home across an athletic field just after an autumn sunset. Looking up, I saw a bright spot of light in the sky, moving too fast to be an airplane. I realized that I was looking at Sputnik, the first artificial satellite, which had been launched a few days before by the Soviet Union.

It's hard to overstate the impact that this 183-pound object up there in near Earth orbit had on the American educational system. The years immediately following World War II had been glory days for American scientific confidence. Some writers had taken to calling the conflict the "Physicists' War," and certainly the success of Allied scientists in developing things like radar, the proximity fuse, and, of course, the atomic bomb had demonstrated for all to see the West's clear dominance in the world of science. A generation of scientists came back from the war ready to propel America into a brilliant future, swaggering and, as one observer put it, writing out of the sides of their pens. During my own student years I was always amazed at how many of my professors—mild-mannered, pleasant men—had played critical roles in wartime research.

My own favorite example of the esteem in which scientists came to be held during and after World War II comes from the autobiography of Luis Alvarez, whose Nobel Prize in 1968 capped a distinguished career in experimental physics. During the war he was at the Radiation Lab at MIT, one of the centers of wartime research, where he worked on developing the proximity fuse. Up until that time, people manning antiaircraft guns would estimate the height of incoming aircraft and then set the fuses on their shells so they would go off after a certain amount of time, with any luck, somewhere near the aircraft. Needless to say, this technique didn't produce many hits. The *proximity fuse* is basically a small radar set in the nose of the shell that ignites the explosive when the shell is near an aircraft. On a test mission on Chesapeake Bay, shells equipped with the scientist's new fuse shot down one drone plane after another, to the amazement of the Navy gunners. For security reasons, the ship was immediately sent out to sea, but a boat was launched to take the scientists back to shore. The commander came down to make sure his charges got off safely. He shook each man's hand, then turned to the officer in charge of the boat and said, "Lieutenant, see that these men are wearing their life vests." It was a mark of respect that stayed with Alvarez his entire life.

But in 1957, thanks to one successful Soviet satellite launch, all that confidence, all that complacency, was gone. Every hour and a half, the "beep beep beep" from Sputnik passing overhead reminded us that our enemy in the cold war had beat us into space. My memories of this period are of a sense of public panic and bewilderment, a similar but much more muted version of what happened after 9/11. The American mood was captured by Alabama Senator Lester Hill in his 1958 testimony to the Senate Committee on Labor and Public Welfare:

> The Soviet Union, which only 40 years ago was a nation of peasants, today is challenging our America in. . . science and technology. . . . The path we choose to pursue may well determine the future not only of western civilization, but freedom and peace for all peoples of the earth. [9]

Americans everywhere wondered how this could have happened. How, so soon after our great triumphs, could we have found ourselves in such an apparently desperate situation?

What followed next was a typical American reaction (over-reaction?) to a threatening situation. The country suddenly went into high gear as far as science education was concerned. Were other countries producing more scientists and engineers than we were? We'll start producing more than they could dream of. Were the Soviets in space? We would beat them to the moon. Money poured into the scientific establishment.

And it bought results. The National Defense Education Act (NDEA), authorized in 1958, pumped billions of dollars into training a new generation of scientists and engineers. In the end, the amount of money spent through NDEA exceeded what had been spent on the Manhattan Project in the first place. In my own case, I was able to use some NDEA money to support graduate students before the program was terminated decades later.

The work of the Physical Science Study Committee (PSSC) was symbolic of the sudden surge of interest in science education. Composed of MIT scientists (mainly physicists) and industry leaders, the committee was actually put together the year before Sputnik. Driven by the sudden post-Sputnik furor, it produced PSSC Physics, an extremely well thought out (though advanced) high school physics course that was soon available in schools all across the country. Indeed, I still encounter the occasional undergraduate who has had this physics course.

What was most striking about PSSC to me when I spent a postdoctoral year at MIT in the late 1960s was the fact that several very senior physicists were still involved with it, people like Jerrold Zacharias, who is usually considered to be its founder. Their participation made it acceptable for lesser physicists to be interested in educational topics, to the point that education was occasionally the topic of conversation at the daily brown bag lunches of the theory group—something that would almost be unheard of today.

Through the 1960s, PSSC concentrated on teacher training, running intensive summer content institutes for high school teachers around the country, and producing classroom materials. In the end, though, it never really caught on—I don't think it ever captured as much as 10% of the high school physics textbook market. My own sense is that its intellectual level was just too high for most students and teachers—a point to which I'll return in a later chapter.

But in any case, the billions spent on science education by President Eisenhower and his successors paid off. Students flocked into new science fields and scientists trained in basic research poured into the research pipeline. By the end of the 1960s, with American footprints on the moon and a seemingly unending string of Nobel Prizes, American science had once again asserted its dominance in the world. Students from all over the world came to American universities for graduate training, as Americans had, only a few decades earlier, gone to Europe.

Having said all this, and with due respect for the huge expansion of the American research enterprise that followed Sputnik, we still have to ask our important question: What about the other 98%? What, in other words, about the students who were not going to pursue careers in science and technology? On this score, the effect of the post-Sputnik science boom was equivocal at best. On the one hand, it did serve to create interest in science education and bring some senior scientists into the field, as discussed above. On the other hand, as New York University physicist Morris Shamos said in his book *The Myth of Scientific Literacy*, throughout this period

> the primary function of formal science education, whether precollege or college, [was] to ensure a steady supply of scientists and science-related professionals. [10]

This was certainly the goal of PSSC, although they included in their target elite the kinds of nonscientists who would be important in American society: future lawyers, bankers, political leaders, and so on. In the words of Jerrold Zacharias, "We had to establish a first-class collection of stuff for the intellectual elite of the country."[11]

AMERICAN EDUCATION AFTER THE POST-SPUTNIK BOOM

In this concentration on the upper end of the educational spectrum, and especially on the recruitment of future scientists and engineers, the post-Sputnik boom set the tone for what was to follow as the twentieth century played out and the twenty-first began. There were a number of landmark events in American education during those years; the publication of the report *A Nation at Risk* in 1983 by the National Commission on Excellence in Education and the passage of No Child Left Behind (NCLB) in 2002 both loom large in retrospect. Both were driven by a perception that there was something seriously amiss in American public education, and both advocated far-reaching reforms. Neither was focused exclusively on science education, although both included science and technology in their purview.

Enough time has passed since 1983 for us to be able to make some sort of evaluation of the effects of *A Nation at Risk*. From my point of view, the most important result of this report, which warned of a nation drowning in a "rising tide of mediocrity," was the placing of education squarely center stage in terms of public attention. When was the last time, for example, that a political aspirant didn't claim that he or she wanted to become "the education president"? In the words of Nobel laureate Glenn Seaborg in 1993, "It is now apparent that the precollege educational crisis and the urgent need for educational reform are broadly perceived as being a top priority."[12] The tightening of high school graduation requirements and college entrance requirements in the last 2 decades of the twentieth century can, I think, be thought of as a direct consequence of the publication of *A Nation at Risk*.

On the other hand, there has been surprisingly little improvement in measurable educational outcomes as a result of these reforms. In essence, apart from some improvement in mathematics, test scores for American students have been approximately flat since the 1970s, so any positive effects the reform may have had have not been so great as to become visible in the (admittedly imperfect) measures provided by standardized testing.

It's much too early to make a similar assessment of No Child Left Behind, although its introduction of national standards, its increased emphasis on basic skills like writing and mathematics, and its program of public accountability has already changed the general atmosphere in the educational system. Nevertheless, we can already see some changes in science education emerging as unintended consequences of NCLB. The increased emphasis on basic skills in the early grades is tending to push nontested subjects like science and history up the educational ladder so that students encounter them later in their careers. It is unclear at this stage, at least to me, whether this will have a positive or negative effect on student achievement in science. On the one hand, having a sound base in language and math will surely help students who had to struggle with the language in grade-level textbooks under the old system. On the other hand, we don't know what putting off the introduction of science will do to students' natural interest in the subject. This is, I think, a question for future research.

But whatever the long-lasting effects of *A Nation at Risk* and No Child Left Behind turn out to be, as far as science education is concerned the rhetoric surrounding them sounds a lot like the familiar post-Sputnik refrain: American students are falling behind students in other countries; we are not producing enough scientists and engineers to keep the economic engines going; unless we immediately make drastic changes the entire educational enterprise will collapse.

Following Bjorn Lomborg in his book *The Skeptical Environmentalist,* I will call this sort of argument *The Litany.*

One amusing feature of The Litany is that over the time the primary foreign competitor—the bête noir of American science, if you will—has changed. In the 1960s, it was the Soviet Union; then in the 1980s it was Japan, followed by Taiwan, Singapore, and the other Asian tigers. Right now my sense is that the next country in line to be bête noir is either China or India, depending on who is doing the writing and whether the primary concern is overall industrial dominance or the outsourcing of American high-tech jobs. My own guess is that China will eventually inherit the bête noir mantle and become America's chief competitor.

I will look at The Litany in a little more detail later, with an eye toward making a realistic assessment of different countries' relative success in producing engineers and scientists. For the moment, however, let me turn to the question I raised earlier: Where, in all of this furor, does the nonscience student fit in? Where is the thinking on scientific literacy?

Physics for Poets and All That

My own personal landmark for the time when the attitude of the scientific community toward teaching nonscientists began to undergo a sea change was the 1972 publication of Robert March's textbook, *Physics for Poets.* March was a respected University of Wisconsin experimental physicist, something of a free spirit in a buttoned-down world. When the book came out, few scientists were thinking seriously about any educational issues other than those involved in producing the next generation of scientists and engineers. The book started to change all that.

First, of course, there was the catchy title. It caught people's attention and made them think about teaching nonscientists. Second, the book was well written and well thought out. It would have been difficult to dismiss it as "watered down," as many scientists at the time were inclined to do. And finally, the book appeared just as the enrollment pressures discussed in Chapter 9 began to make themselves known, so that university science departments were open to the idea of expanding their student numbers. One by one, departments that had held back until then began to add another introductory course to their portfolio. In physics, for example, the new course aimed at nonscientists was added to the existing introductory array of algebra- and calculus-based courses (the former is intended primarily for pre-medical students and the latter for engineers and scientists). Today, it would be a rare science department that wouldn't have some sort of "Poets" course on its books.

I was at the University of Virginia while all of this was going on. The physics department was a typical, somewhat conservative, operation, but they allowed me to try introducing a "Poets" course. I can still remember the day when the enrollment passed 300. The back row of the lecture hall was filled with my colleagues, who just couldn't believe that that many students wanted to take a physics course. (To be realistic, since this was the first "Poets" course at the university, students undoubtedly saw it as the easiest way through the science requirement, a theory that was supported by the nickname they gave it—Football Physics.)

This is not to say, however, that the movement begun by *Physics for Poets* was really aimed at producing scientific literacy. Every course introduced during this period was based in a specific discipline— Geology for Poets, Chemistry for Poets, and so on. The intent was clearly to provide nonspecialists with a deeper appreciation of a specific discipline, rather than to give nonscientists a broad background in the sciences. Even with all of these courses in place, it was possible for a student to fulfill the university science requirement and never hear terms like *DNA* or *semiconductor* uttered in a classroom.

I would like to think that the 1995 publication of a textbook titled *The Sciences: An Integrated Approach* by Robert Hazen and myself will mark a similar milestone in the way we teach science at the university level. Based on the Great Ideas approach that I discuss in Chapter 12, this book marks a radically different approach to nonmajor education. For the first time, it was possible to find a textbook for a course that gave a grand overview of all the sciences, rather than a presentation of just one field. It is much too early to make an assessment of the impact that this book has had, but there are some encouraging signs. The text is now in its fifth edition, and courses based on its approach have been instituted at more than 200 colleges and universities around the country. Recently, it was paid the ultimate compliment of having other authors and publishers bring out competing textbooks—a sure sign that interest in the subject has reached a critical point.

But even with the addition of "Poets" and "Integrated Sciences" courses in universities across the country, the attention of the scientific community largely remains fixed on recruiting new scientists and engineers, with education for other professionals coming next and education for the public, as always, dead last. This is an understandable, though somewhat discouraging, state of affairs. In an attempt to loosen the grip of this attitude on my colleagues, I would like to close this chapter by asking a simple question: Are we really falling behind in the production of future scientists and engineers?

Are We Really Behind?

That American educators should feel insecure about the results of their work in science is, on the face of it, somewhat hard to understand. With few exceptions, American science still sets the world standard and, as pointed out above, American universities are still the educational venue of choice for students around the world, especially those seeking degrees in science and engineering. Yet when you look at the popular media, all you hear is The Litany. How, in the face of our manifest success, can people seriously argue that the system is failing?

One reason, of course, is the poor performance of American high school students on international standardized tests in science and mathematics. Another comes from comparisons of graduation rates for scientists and engineers in different countries. A typical argument might go something like this: In 2004 (the last year for which there are data as of this writing), the United States produced about 70,000 engineering graduates, while in that same year India produced 350,000 and China produced 600,000. Clearly, other countries are doing better than we are in this vital educational task.

Because I lived through the post-Sputnik hysteria, you will excuse me if I look at this current version of The Litany with a jaundiced eye. I am indebted to a group of scholars in the Master of Engineering Management program at Duke University, under the guidance of Gary Gereffi and Vivek Wadhwa, for taking a close look at these statistics and restoring a semblance of rationality to the discussion of the production of technically trained professionals. What this group did, in essence, was to look in some detail at what counts as an engineering degree in India and China, and then try to make an apples-to-apples

Table 8.1. Comparison of 4-Year Baccalaureate Degrees in Engineering in 2004.

Country	Number of degrees	Degrees per 1 million population
China	351,537	271.1
India	112,000	103.7
United States	137,437	468.3

comparison between those countries and the United States. As often happens, when you look closely at where the numbers in the headlines come from, the situation seems a lot less serious than it first appears.

Let's start with China, where the Duke researchers contacted the Ministry of Education for their data. According to the ministry, in 2004 some 644,106 engineering degrees were awarded. Of these, 351,537—a little more than half—were from 4-year baccalaureate programs, and the rest were from what are called "short cycle" programs. These are typically 2–3-year programs, roughly analogous to the associate degree given at American community colleges. But even here there is a problem, since the ministry apparently simply collates graduation reports from the provinces, without imposing a uniform definition of what is meant by an engineering degree. Consequently, some of the degrees included in the total probably were given to people like auto mechanics and heating technicians.

A similar situation applies in India. Because the Indian government does not collect graduation data systematically (the last officially available graduation data are from 1993), the Duke researchers went to the National Association of Software and Service Companies (NASSCOM), which serves as the de facto collector of this sort of data in the country. For technical reasons, the NASSCOM numbers for 2004 were projections rather than hard data, but they tell a story similar to China's. They estimate that in 2004 India produced 215,000 engineering degrees (not the 350,000 of the headlines), of which 112,000 (again, a bit more than half) were from 4-year programs.

To get Table 8.1, the Duke researchers put together these numbers with results from the National Center for Education Statistics in the United States (137,437 engineering degrees from 4-year programs), and used official population figures for the countries to determine how many degrees were awarded per 1 million people. Looking at the statistics this way, comparing apples to apples, the gap in educational attainment seems to vanish. In absolute numbers, we're clearly in the game with our two chief rivals. The per capita production of degrees is, perhaps, the more important number, since the need for engineers should go up with population—after all, *someone* has to design all those extra freeway interchanges. Here the American advantage is pronounced.

In a rapidly globalizing world, Americans can't afford to be complacent about our educational system, nor about our ability to compete internationally in the global arena. We can always do better than we're doing right now. On the other hand, as the above table shows, we're not nearly as bad off as The Litany would have it.

CHAPTER 9

Apportioning the Blame: How We Got to Where We Are

W e find ourselves in a situation where scientific literacy is important, but where for some reason our schools are not producing scientifically literate graduates. Despite the recent progress in this area (documented in Chapter 6), I have a sense that somewhere along the line America, a country built on new science and technological innovation, seems to have gotten off track. Before we can talk about how to improve things, we ought to have a clear notion in mind of how this state of affairs came to be.

In assigning blame, it is important to distinguish between two kinds of causes: ultimate and proximate. As with any major problem, the ultimate cause of scientific illiteracy can be laid on society as a whole, and particularly on popular attitudes about science. This is a mushy and unsatisfying statement, more a counsel of despair and an admission of defeat than anything else. It certainly suggests little in the way of remedy, so I will not dwell long upon it. I am much more interested in finding out why, given widespread scientific illiteracy, our educational institutions have not responded to the problem, regardless of its ultimate cause. About this I will have a great deal to say, for the fact of the matter is that the educational establishment must be galvanized into action if the next generation of American students is not to turn out as scientifically illiterate as the present one.

In what follows, I will talk about the educational progression from elementary school to the university. In some areas, such as elementary school education, I can pretend to no more expertise than would the average concerned adult and parent, but I have had some experience with the middle school curriculum. I can speak from a lifetime of sad experience about the problems involving university education. In the early stages of education, I will argue that the attitude that "we teach children, not subjects," which seems to be pervasive in the system, discourages serious study of science. Later on, in high school and college, a rather insidious alliance between scientists (who want to teach only the highest-level students), humanists (who don't want to learn science at all), and the students themselves (who want to get through as easily as possible) militates against education in scientific literacy.

Although the tone of this chapter may seem somewhat negative, it should not be seen as a jeremiad. Even as I write these words, hundreds of very talented and highly motivated people are working on finding ways to improve science education in this country. Federal agencies are pouring money and effort into this area as well. For example, I am sitting on a committee of the National Academy of Science charged with reviewing NASA programs in K–12 education. This agency spends tens of millions of dollars annually on this venture (as well as considerably more on supporting graduate students in the sciences and engineering). Other agencies achieve the same end by requiring recipients of research grants to allocate some of their money to educational outreach programs. Consequently, the picture is not all dark, despite the severity of the problems I want to discuss.

ELEMENTARY AND MIDDLE SCHOOL

The general consensus among educators is that what I call the great turnoff to science occurs sometime around the middle school or junior high years. Normal curiosity about the world seems to turn into disdain for, and perhaps even a fear of, things scientific. Many people are studying how and why this happens; peer pressure, parental attitudes, and the perceived difficulty of doing science have all been put, among others, on a list of the usual suspects.

In one sense, it is hard to see why there should be a special problem with science education. Children are naturally curious. A 5-year-old can hunch over and watch a butterfly or a beetle for periods far exceeding a child's normal attention span. A 10-year-old at the beach may spend an afternoon accumulating a treasure trove of shells and rocks, a dripping mass that the parents have to transport home. This curiosity is the raw material on which science feeds, the stuff that, at bottom, motivates most adult scientific research. In this sense, young children are naturally interested in science. Somehow that all changes in middle school.

A personal experience illustrates one major problem in the teaching of science at this level. The son of a colleague and my oldest daughter happen to be the same age, so both were taking the same science course in their schools. I should mention that because of the peculiar geography of the D.C. region, my colleague's son was at a school in Maryland, while my daughter was in Virginia. Both schools, however, were located in affluent suburbs full of engaged, highly educated parents, and both schools are routinely listed in the top 100 in the nation in *Newsweek*'s annual rankings.

In any case, the curriculum came around to the section on meteorology. Weather is one of those subjects in which children are intrinsically interested—how many times have parents been asked what the clouds are or why it rains? My colleague's son, however, was handed a list of more than 20 meteorological instruments to memorize. It was not, mind you, a list of words with explanations of why they were important; it was simply a list to be memorized.

Needless to say, my colleague was furious. I got such an earful that at dinner that night I asked my daughter if she knew what a *swinging hygrometer* was (this was one of the items on the list).

"Oh yeah," she said. "That's one of those things you take outside and swing around to measure the humidity in the air." Her body language made it clear that her class had actually gone outside and used the instrument, rather than just memorizing its name. I was relieved to hear this—especially after all the property taxes I'd paid! But this episode certainly illustrates how easy it is to take an interesting piece of science and turn it into something deadly dull.

Unfortunately, the experience of my colleague's son is not an isolated incident. It has been estimated that in eighth grade a student is likely to encounter more vocabulary words in his or her science class than in English. It's not so much that the list he was given included things that only a professional meteorologist would know about, but that what should have been a vital, engaging, hands-on subject was turned into a dry exercise in rote memorization. Too many students, I fear, think that all science is like this—that in college you memorize 100 instruments and those who go on to a Ph.D. memorize a thousand. Is it any wonder that students exposed to this sort of teaching wind up bailing out of science courses as soon as they can?

Until recently, I would have placed much of the blame for poor teaching of science in the middle schools squarely at the door of university schools of education. I still hear occasional echoes of the old "we teach children, not subjects" in professional circles, although the idea that a teacher has to master subject matter (in addition to teaching methods) is more widespread now than it once was. Part of this bias against subject matter was the notion that the goal of education, at least in the elementary and middle schools, was to enhance self-esteem, to make the students feel good about themselves. In such a situation, difficult subjects like science tend to be downplayed, since they involve a risk of failure and consequent loss of self-esteem.

I wish I could say that this attitude in the schools has faded since the advent of No Child Left Behind, but my discussions with teachers who are in the trenches lead me to suspect otherwise. The idea that teachers are to devise a strategy to let every child succeed sounds good on paper, but it assumes that all children really want to succeed in academics—a doubtful proposition if there ever was one. It also assumes that all subjects are equally easy to master, something that is surely not true for science and math. Thus all too often we have a situation in which children are to be protected from failure in order to enhance their emotional well-being, without at the same time demanding sufficient effort from them to allow them to truly master their subjects.

The ultimate example of the consequences of this sort of attitude were displayed in a report titled *Science and Math Education in a Global Age* put out by the Asia Society in 2006. In side-by-side graphs they exhibited the results of two studies. One showed the percentage of eighth-grade students in different countries who achieved an advanced score on a standard math achievement test. The winners (no surprise here) were the Asian countries—Singapore, Taiwan, South Korea, and so on—with percentages ranging from 24% to 44%. American students came in at 7%. The next graph showed the results for the same students on a test of student self-confidence. Here Americans led the field, with 56% of our students confident that they were good in math. (For reference, South Korean children, 35% of whom achieved advanced scores on the actual tests, scored only 20% in self-confidence.)

Unfortunately, I have met all too many educators (particularly among elementary and middle school administrators) who seem to think that this process of building confidence on air instead of solid achievement is actually the proper goal of the school system. One reason I find this attitude so discouraging is that I recently came across an essay I wrote in the 1980s, in which I commented on a similar study that had results distressingly similar to those in the Asia Society report. In that study, the South Korean students finished first in math achievement, the Americans last. When it came to evaluating the statement "I am good at mathematics," fully 68% of the American students responded to this in the affirmative, indicating that their education had, indeed, made them "feel good about themselves." In response to the same question, only 24% of the South Koreans answered yes.

Fortunately, there are signs that content may be coming back into fashion in the educational establishment, and most states now require that teachers have a bachelor's degree in a specific discipline before they go on to advanced study in teaching itself. Before this welcome change began, teachers would routinely be expected to teach scientific subjects over which they had little mastery. In such a situation, any teacher (the author included) would feel uncomfortable, and it is not hard to see that it would be hard to convey an interest in (much less a love for) science under those circumstances.

It seems to me that this situation is particularly vexing at the elementary school level. The level of scientific knowledge needed to teach a student at this level isn't all that high; after all, how sophisticated can a 6-year-old's questions be? What is needed, I would argue, is for the teachers to feel comfortable with the material, so that they don't convey a sense of unease (or, worse yet, fear) to the students. What elementary school teachers really need is the matrix of knowledge I have called scientific literacy. I can remember, for example, that when my daughters were in elementary school it became clear that their teachers were really uncomfortable teaching the units on astronomy. I offered the teachers a couple of informal afternoon tutorials on the subject—nothing technical, just the kind of thing I would later incorporate into scientific literacy courses, just enough to make them comfortable with the subject. They told me that the experience had improved their teaching immeasurably.

I often think of this experience when I think about the preparation of elementary school teachers in our schools of education. I wonder how much general scientific literacy would increase if all of these future teachers were required to take a scientific literacy course in addition to courses in methods.

Finally, we should realize that science really doesn't have to be odd man out in elementary and middle schools. There is, after all, nothing inherently antiscientific about adolescence. Young people in Europe and Japan seem to be able to deal with amounts of scientific knowledge that are little short of prodigious by American standards. Nor is adolescence a time when young people necessarily shy away from difficult endeavors—how many musicians and athletes first became serious about their careers during this period in their lives? It seems to me that the childhood interest in nature should serve as a precious capital that the students can use to tide them over the tough period when learning science starts to become more difficult than just watching beetles or butterflies.

Having said this, I have to add that some recent experiences with textbook development (not to mention getting married to an extremely skilled middle school teacher) have given me a new appreciation of the aspects of science education that do not come from science itself. My flash of enlightenment came at a conference being held at the headquarters of McDougal-Littel publishers in Evanston, just outside of Chicago. The company has a long history of publishing elementary and middle school texts in subjects like literature and history, and was branching out into science. I was at the meeting as the "content guy," and, like most scientists, felt that getting the content right was all there was to it. Looking around the table, I realized that everyone was there because he or she brought expertise in some subject—curriculum, reading level, English as a second language, and so on. It suddenly struck me that getting the content right was just one piece, albeit an important piece, of a much larger picture. In order to succeed, we had to get everything right, not just the content. I have to say that that experience gave me a much deeper appreciation of the problems involved in presecondary teaching than I had had up to that time.

One thing that this experience taught me was that the popular games of Blame the Schools and Blame the Teachers have a limited potential to help with the problem of scientific literacy. As will become clear as we work our way up the educational ladder in this chapter, there is plenty of blame to go around. The standard attitude I encounter among the minority of my university colleagues who care about presecondary education is basically, "Well, here's the science, and that's the end of my responsibility." At the very least, scientists have to become more aware of the incredible complexity of the non-scientific problems involved in elementary and middle school education, and perhaps be a little less self-righteous. Frankly, I don't know any university scientist who would be able to last for a day with classes of 13-year-olds; I know that I wouldn't.

When I want to make this point during a public lecture, I often point out that for me, speaking to a large audience is easy. I do it all the time. I then say, quite truthfully, that the only time I was ever terrified before a public appearance was when I had to talk to my daughter's third-grade class about magnetism.

Enough said.

HIGH SCHOOL

There doesn't seem to be any single cause for the great turnoff, but it seems to me that if the schools can't help, at least they should do no harm. Unfortunately, from what evidence we have, we cannot conclude that they are playing even this neutral role at this time. The high school years in America are a continuation of the turning away from science that starts earlier. Only 23% of American high school students take as many as 3 years each of math and science, and fewer than 10% take a course in physics. When students do take courses, it is often to satisfy college entrance requirements rather than those imposed by the state or local school boards.

But oddly enough, this emphasis on college entrance requirements may work to the detriment of science education in other, more subtle, ways. Serious science courses are perceived (rightly, I think) to be difficult for many students. Thus students who take more science courses than absolutely necessary are, in effect, risking their grade point average and putting their college admission at risk. So while college admission policies may play a role in inducing secondary school students to take science courses, they may also play a role in keeping them away. But college entrance or no, the fact remains that by the time American students get to high school, they are already conditioned to avoid science whenever possible, and high schools do little to change this situation.

It is my sense that, at least in well-funded urban and suburban school districts, the teaching of science to those students who attempt courses in the secondary schools is in pretty good shape. High school teachers tend to have taken at least a minimal number of courses in the disciplines they teach, and have to be certified by subject area. In addition, high school science education has benefited from the fact that university faculties often look back down the pipeline to the high schools and, in an exercise in enlightened self-interest, involve themselves in high school science education. The PSSC physics course I described in Chapter 8 is widely used for the better college-bound students. Today, there is some evidence that a new wave of concern may once again be turning the attention of university scientists to the high schools.

Lest a false sense of euphoria develop, however, we should point out that almost all of the attention being paid to science education these days is aimed at producing future scientists and engineers, not the more general problem of scientific literacy. In fact, for students outside the ranks of college-bound science majors, the forces at work in the high schools seem to be pretty solidly arrayed against the acquisition of scientific literacy. Since the middle schools do nothing to prevent the great turnoff, students often enter high school wanting to avoid science as much as possible. They are then confronted with fragmented, compartmentalized science offerings and asked to pick something to study to "satisfy the science requirement." Forced to choose between courses labeled Physics, Biology, General Science, Earth Science, and the like, they make what is, given the way the system works, the most intelligent choice. They take the minimum number of courses required from among those that have the reputation for being easiest—a pattern that repeats once they get to college.

The fragmentation of science into mutually exclusive domains (physics, chemistry, biology, astronomy, and so on), coupled with the idea that students need to study only a few of them to learn science is a phenomenon that first surfaces in high school. It reappears, in more virulent form, at the university level. I will discuss the intellectual vacuity of this idea later, but for the moment I simply note that students who have chosen to take their required course in earth sciences are unlikely to be able to read an article about genetic engineering intelligently, while students who choose biology may be at a loss when it comes to understanding an earthquake in San Francisco.

In the best of all possible worlds, we would expect that high school graduates would be scientifically literate—you shouldn't have to have a Ph.D. to read the newspaper, after all. But given the system of compartmentalized electives, the best we can expect is that the student not be turned off to science in high school. There is almost no way for an American high school graduate to acquire the kind of broad knowledge that goes into making up the requisite matrix of background knowledge. In fact, in view of the system of elementary and secondary education we have just analyzed, it's a wonder anyone in the country is scientifically literate.

UNIVERSITY

What about the universities? American higher education is, quite rightly, regarded as being the best in the world, particularly at the graduate level. The student arriving at the university already set on a career in science or engineering will find resources and faculty time available, and the amount of each grows the farther along the educational pipeline the student progresses. But, as was the case in the high schools, no such attention is paid to the problem of educating those not planning to be scientists.

As a university scientist, I have had ample opportunity to see how my colleagues react to the problem of teaching nonscientists, and consequently I have evolved some definite (if unflattering) notions about why universities do so little to produce a scientifically literate public. In my view there are two central issues that have to be addressed when we look at the fate of scientific literacy: One is the extraordinarily low state into which the teaching of all subjects (including science) has fallen in our universities; the other is the strange alliance that has developed to deal with the place of science in the general curriculum.

There is a common, if idealistic, view of universities in which they are seen as the repositories of human knowledge, ivy-covered bastions of learning where the accumulated wisdom of our culture is passed on to the next generation. I confess that I find this ideal compelling; I would like our universities to approach it. But if you think that the great universities of the twenty-first century are places where undergraduates discuss the meaning of life on strolls through the woods with gray-haired professors, you obviously haven't spent much time around universities lately. I am put in mind of a dinner party I attended years ago at a major northeastern university. One guest (not affiliated with the university) made it clear that he thought that the education of undergraduates was the highest priority at this particular institution. After a moment's embarrassed silence from the assembled faculty, a senior member of the psychology department smiled sadly and said, "How touching."

What generated that embarrassed silence and the somewhat cynical remark that followed was the realization that in today's university teaching, particularly the teaching of undergraduates, is simply not considered very important. In fact, as seen from the point of view of the modern professoriat, the function of a university is no longer to teach but to provide support for faculty to carry out research.

To understand what we mean by this statement, you have to realize that there are actually two universities that exist, albeit uneasily, on every campus. One is the visible university—the buildings, students, faculty, and staff. It is the visible university that we most often talk about, whether discussing educational policy, the fortunes of the football team, or the latest addition to the high-tech industrial park down the road. From an intellectual point of view, the visible university is organized horizontally, with different branches of knowledge side by side.

But there is an invisible university, too, one made up of a large number of parallel but unrelated invisible colleges. These invisible institutions are made up of scholars whose work is concentrated in a given area. High-energy physicists, for example, make up one invisible college, molecular biologists another, economic theorists a third. Smaller invisible colleges are grouped together to make larger ones; a half dozen subspecialties can be combined to make the invisible college of physicists, for example. Thus the invisible colleges are roughly equivalent to university departments—physics, French, English, and so on. Together, they make up what I am calling the invisible university. Unlike its counterpart, the invisible university is organized vertically, by subject matter.

Every faculty member owes allegiance to both universities. This has, I suppose, been the situation for scholars since universities were first created in the late Middle Ages. There has always been a certain amount of tension between these loyalties, because the two universities make different demands. The visible university is concerned with the education of the students who pass through its doors, as well as with everything from athletics to curriculum reform. The invisible university, on the other hand, is concerned with one thing only—scholarly research. In a rough sense, you can say that the demands of the two universities pit the local against the international, teaching against research.

Living with these schizophrenic demands is the great challenge of academic life. To see why, consider the fact that a university's reputation depends primarily on the reputation of individual faculty members. All the good things of modern academe—federal money to support research and graduate studies, prominent participation in national and international programs, favorable mention in prestigious publications—flow from reputation. New universities on the make and old universities trying to maintain their status all deal in this coin, so that high-prestige faculty are courted and wooed with some of the same fervor that we see expended on first-round draft picks in the National Football League.

Since World War II there has been a fundamental shift in the relation between the two universities as far as their ability to affect reputation is concerned, with the balance of power now firmly on the side of the invisible. The judgment about the reputation of individual faculty members is now made almost solely on the basis of standing in the invisible university. The term *reputation* or *scholarly standing* may seem to be a vague concept, one that is rather difficult to define with precision. Nevertheless, members of the invisible college, if asked, will usually give remarkably similar assessments of another member, just as members of an athletic team will agree on which of their teammates exhibits *leadership*, even if they can't give a precise definition of the term.

However the invisible colleges arrive at the assessment, one thing is clear: The assessment is all that matters in the modern university. In order to build their reputations in the invisible university, faculty members have to carry out and publish research—the familiar "publish or perish." The means to do research—federal research grants, fellowships, and the like—are typically awarded by a process known as peer review. In peer review, members of the invisible university (usually anonymous) pass on the merits of the research proposals, in effect, delivering the judgment of the invisible university on the applicant. The same process governs what will be published in the prestigious scholarly journals. The research a professor does and where—or even whether—it is published depends on the judgment of the invisible university.

The invisible universities began to grow in importance in the late nineteenth century, about the time that universities came to be centers of research and to be organized in departments. Since then, they have operated more or less as outlined, concerning themselves with research exclusively. Indeed, they can operate in no other way: how is a colleague in London supposed to know about, much less evaluate, the teaching of a professor in California? Until the great research boom following World War II, however, the evaluation of the invisible university was always balanced by a countervailing evaluation from the visible university. This rating, much more broadly based, included items such as teaching and service to the university and community. Today, however, the values of the invisible university dominate the decisions of its visible counterpart. The emphasis on reputation and research carries over even into the visible university's internal decisions about promotion and status. This process is typically carried out by something called a P&T (promotion and tenure) Committee. The first thing the committee does is to "go for letters"—a shorthand phrase that denotes asking selected members of the invisible university for an anonymous assessment of the candidate's standing with his or her peers. Without positive comments in those letters, promotion comes very hard in today's university.

The response of the faculty to this state of affairs has been rational, if unfortunate. If promotion, tenure, and salary are going to depend almost exclusively on scholarly standing, then scholarly standing is what will be sought. And since scholarly standing is determined by the invisible university, those things that the invisible university values will be done. What could make more sense than this response?

Unfortunately, the conduct most likely to benefit the individual faculty member is precisely the conduct least likely to benefit the education of undergraduates. In fact, over the last 50 years there has been a wholesale flight from teaching in our university faculties. As any economist or psychologist could have predicted, when a system is set up to reward one kind of behavior and punish another, the rewarded behavior will eventually dominate. In this case, the rewards of academic life go to those who succeed in research, so that's where most academic effort is focused. In the summer of 2007, the Harvard University faculty released a report on faculty governance at that institution, aimed as countering the overemphasis on research discussed here. One physicist commented about the current situation as follows: "I wish my colleagues at Harvard good luck in implementing the recommendations in the new report, but, frankly, I'm not wildly optimistic about their chances of success." [1]

It's hard to overestimate the negative effects that this emphasis on research has had on teaching at the university level. At one level, it is a matter of simple arithmetic. There are only 24 hours in a day, and any time that an academic devotes to improving teaching is necessarily time that cannot be spent doing research. This fact is openly acknowledged in most universities. In the early 1970s, for example, a new associate professor at a major university was given a bit of friendly advice by his department chairman: "If you spend more than 10% of your time on teaching," he was told, "you'll be hurting your chances for promotion to full professor." To be perfectly honest, I would give roughly the same advice to new faculty entering almost any university today, although I might up the percentage from 10% to 20%. This isn't cynicism, but a simple appraisal of how the system works.

The reward system we have described operates in full force in science departments, perhaps more so than elsewhere. In fact, during the last couple of decades a further refinement has emerged in which faculty members are not judged by the quality of their research or the number of publications, but by the dollar amount of their research grants. In many science departments, in other words, it is now possible to publish and perish.

Most universities pay at least lip service to the importance of teaching, but when the chips are down, the faculty knows that it's research that counts. In all of my experience in academe, I have never heard of a professor being promoted on the basis of teaching excellence if his or her research record wasn't up to snuff. I have, however, seen numerous cases where the opposite held true, where a good researcher was promoted despite the fact that he or she was an unmitigated disaster in the classroom. I can recall an exchange that took place at the end of a long and rancorous meeting of a promotions and tenure committee at a major university. One member stated that the university demanded excellence in both scholarship and teaching.

Another member, a historian whose low tolerance for pomposity had been lowered even further by the long meeting, shot back, "That's [expletive deleted]. We demand that they be excellent in research and that their teaching not stink up the house." That seems to me to be a fair, if blunt, assessment of the relative value placed on the two main missions of the modern university.

If the distortion of values at promotion time were the only effect of the overemphasis on research standing, it would be a problem, but not necessarily a disaster, for university education. Far more serious is the contemptuous attitude toward teaching that the situation encourages. All too often, teaching is regarded not merely as a necessary evil, a regrettable diversion of time from research, but as an activity in which success is somehow de facto evidence of unsuitability for an academic career. Deep in the recesses of the Faculty Club, one often hears dark murmurings about colleagues whose courses are oversubscribed by enthusiastic students. The suspicion is that the courses must be attractive because they're graded easily and have little intellectual content.

There is actually a little bit of university politics tied up in this attitude, so I'd better explain. In a typical university, one strong influence on the size of an academic department is the number of students enrolled in its classes. This number is measured in something called an FTE (full-time equivalent). The formula that relates students to FTE is complicated and varies a little from place to place, but typically 100 lower-level undergraduates, 30 upper-level undergraduates, or 10–15 graduate students would justify one faculty position. (If you're wondering how a university can value research above all else but apportion positions according to enrollments, all I can say is "Welcome to the wonderful world of academe!")

When you go to the dean to try to get a faculty slot to hire the hotshot researcher du jour, one thing the dean will look at is the number of FTE justified by your enrollments. Since science departments typically have small classes, they often feel themselves at a disadvantage in this kind of situation. Thus a scientist who attracts large numbers of students is valued because those students will allow the research function of the department to go forward. Scientists being human, this means that that person will be regarded with a certain amount of suspicion (he or she must be doing something "unsound") as well as plain old envy.

Actually, large enrollments are not always necessarily prized by academic departments, in spite of their contribution to the departmental FTE. When a well-known scholar (not a scientist, but a brilliant lecturer) at a major midwestern university found enrollments in his course passing the 300 mark, he approached his department chairman about getting extra secretarial help to keep the class rolls. The reply: "Why don't you mess things up a little so the enrollment goes down?"

Unfortunately, it's almost impossible to overestimate the low esteem into which teaching has fallen in the modern university. One of the most prized rewards that any university administration can give a scholar is a light teaching load. In too many prestigious universities undergraduates rarely see the famous faculty, who, if they teach at all, teach only advanced graduate seminars directly related to their research topics. Just ask any undergraduate to remember the last time he or she saw a Nobel laureate. One hallmark of success in academe is to teach less than your colleagues.

Two anecdotes illustrate this point. A noted physicist (who later received a Nobel Prize) at one institution spent so much time supervising experiments around the world that he was actually reckoned to be on travel status on those rare occasions when he returned to home base to teach his graduate seminar!

At another university, a senior faculty member who had openly expressed discontent with the overemphasis on research and low status of teaching at his home university and was entertaining an offer from another institution had his administration put together a counteroffer that had, as its centerpiece, a significantly lowered teaching load!

The general decline in university teaching has affected all fields of study, of course, but has produced somewhat mixed results in the sciences. The reason is that many courses in science departments are devoted to the training of future scientists, and this is something that scientists tend to take pretty seriously. It is not unusual, therefore, to find introductory courses intended for science majors taught by experienced senior faculty; in fact, this is an almost universal phenomenon at top-ranked physics departments. Teaching this sort of course is considered an obligation that goes along with seniority, like serving on committees. Think of it in terms of noblesse oblige. Although there's always room for improvement, it is my impression that the education of future scientists in our universities is in pretty good shape.

When it comes to teaching for nonprofessionals, however, the barriers faced by university scientists are daunting indeed. Activities designed to raise the country's stock of scientific literacy do little to add to anyone's research; they will not help you publish; they are not normally considered in the noblesse oblige category of "real" science courses; and they certainly don't bring in research money. Given the current situation, it would be foolhardy indeed for faculty members to become involved in dealing with scientific literacy.

This state of affairs is particularly damaging in the sciences. If students aren't attracted into science courses by good teaching, the consequences of the great turnoff guarantee that they won't come of their own accord. And although you might think that universities would try to produce scientifically literate graduates regardless of the state of the student body at matriculation, they do not. For the fact of the matter is that an unholy (if unconscious) alliance has arisen lately among scientists, humanists, and students. This triad has produced a system in which it is possible for a student to graduate from a university and still be unaware of the most basic facts about the physical world in which we live and the technology that shapes our lives. In fact, I have seen studies that indicate that fully a third of American undergraduates can obtain a bachelor's degree *without ever taking a science course at all!*

As a scientist, I am most aware of the role that I and my colleagues have played in this sorry story. Scientists often speak of science education these days, but if you listen carefully, you find that they are really discussing the problem of finding and training the next generation of scientists and engineers, or of attracting students into science. As we have already pointed out, this is a pressing problem, but it has little to do with scientific literacy. The fact of the matter is that in the training of a scientist, the same subject matter is encountered repeatedly, with each succeeding presentation being more mathematical and more complex than the last. In physics, for example, one studies basic electricity and magnetism (with minimal use of calculus) during the sophomore year, at the intermediate level (with vector calculus) in the junior year, and then meets it again in all its glory in one last go-around during the first year of graduate school. To many

scientists, only the last of these is "the real stuff," and the two prelimi-
naries are tolerated because their aim is to produce a student who has
mastered the subject in all its complexity.

To the average scientist, teaching the central ideas of our trade
without their full complement of mathematical panoply smacks
of heresy. *Watered down* is the politest term normally used to describe it. This is contrasted to teaching *real science* (that's actually
the term used behind closed doors). The attitude seems to be that
unless science is taught with the goal of producing future scientists in
mind—miniature copies of ourselves—it is somehow unworthy. Like
a mystic priesthood, many scientists feel that our greatest secret—the
fact that the important ideas in science are simple—should be closely
guarded. Indeed, I can remember a senior theoretical physicist advancing this argument to justify his unwillingness to cooperate with
a television crew that wanted to produce a film for the Public Broad-
casting System.

Even when science departments bite the bullet and offer courses for
nonprofessionals, university organization guarantees that these courses
will be narrowly departmental. Thus you have the Physics for Poets
phenomenon, survey courses that satisfy the science requirement by
providing instruction in one branch of science. Like their high school
counterparts, these compartmentalized courses produce students who
are scientifically illiterate in all areas of science except one—students
who know about geology but not astronomy, about biology but not
physics. Needless to say, this sort of compartmentalization does not
turn out students who can read the newspaper with understanding.

Turning from the sciences to the humanities, we find a very different
situation. You might expect humanists to provide the loudest chorus
of protest at having been denied an important part of their intellectual
birthright by their scientific brethren. In fact, you find nothing of the
kind. In some literary circles ignorance of things scientific is not only
accepted, but seems to carry a kind of perverse cachet. As I pointed
out in Chapter 4, if an engineer has never read Shakespeare, he or
she is rightly considered ill educated, yet no such judgment is made
about a professor of English who has never read Darwin. Too often
we encounter a smug feeling that if science is just ignored, it will go
away and leave the rest of the world in a state of tweedy comfort.

Finally, we come to the third member of the alliance—the students. As they did in high school, college students tend to avoid sci-

ence where possible, reasoning that such courses are tough and that attempting them is likely to endanger the grade point average (or at least involve unnecessary work). If an enlightened faculty forces them to take *some* science, a search commences for the least demanding item in the catalogue.

So the scientists say that science for the general public is not worth teaching, humanists say it isn't worth studying, the students say it's too hard, and nobody wants to teach the courses anyway. Is it any wonder that a headline about genetic engineering scares away so many newspaper readers?

THE ROBINSON PROFESSORS: AN ENCOURAGING DEVELOPMENT

Having gone through this discouraging discussion of the current state of American education, I have to close the chapter on a more encouraging note by talking about a program at my own university. George Mason University is part of the Virginia state system. Located in the suburbs of Washington D.C., it has about 32,000 students— an ethnically diverse student body and a large body of adult learners along with the traditional 18–22-year-olds. In the 1980s, Clarence J. Robinson, a businessman who was also the first chairman of the university's Board of Visitors (trustees), left money to fund more than a dozen chaired professors. George Johnson, then president of the university, asked himself a simple question: "What is the biggest problem in universities today?" His answer: the fact that senior faculty don't teach undergraduates.

His solution to the problem was simple and straightforward. He announced that the endowed chairs would be filled by faculty who had (1) already established themselves in their field of scholarship, and (2) had a clear track record of interest and success in undergraduate education. The areas of expertise of the group he assembled ranged from art history to molecular biology, and I have been privileged to serve in it as Robinson Professor of Physics. This kind of end run around the standard research-based faculty recruitment system represents one way to deal with the problem I have outlined in this chapter, one that I know works from personal experience. I'm sure there are others as well.

CHAPTER 10

The Goals of Science Education

Throughout this book I have been stressing scientific literacy as the main goal of education for nonscientists, and I have been explicit in defining what I mean by that term. At this point, however, I have to point out the fact that many other people have different definitions of the goal of science education for the general population. Since the way we define our goals determines the kind of educational system we will devise, we need to consider these alternate ways of approaching the problem. In the end, I will argue that a minimalist definition like mine not only leads to the only feasible educational scheme aimed at producing scientific literacy, but is profoundly in tune with what has been going on in the sciences over the last several decades

THE GOAL OF SCIENTIFIC LITERACY

For reference, the operational definition of scientific literacy I gave in Chapter 2 was this:

Scientific literacy is the matrix of knowledge needed to understand enough about the physical universe to deal with issues that come across our horizon, in the news or elsewhere.

This definition is based squarely on considerations of the way that average citizens actually use science. Just as they need to know enough economics to read an article about tax legislation and enough law to read about a case pending before the Supreme Court, they need to know, for example, what embryonic stem cells are in order to come to an informed opinion about the various moral and ethical issues surrounding them. This is what I called the argument from civics in Chapter 3. In Chapters 4 and 5, I argued that science is an essential part of the culture in which we live and that a basic understanding of science can add to our aesthetic experience of the world around us.

One essential feature of all of these arguments was that the type of knowledge needed by average citizens is quite different from that required of practicing scientists or engineers. It does not include the ability to do science or (as I shall argue more fully below) the ability to manipulate mathematical equations. I will lay out a detailed program for scientific literacy in Chapter 12, but for the moment I can say that it is based on a general understanding of the basic principles by which the physical world operates. Thus scientifically literate people would know that energy can be neither created nor destroyed, but can be transferred from one form to another—from solar radiation to electrical current, for example. They would not, however, necessarily be able to analyze the suitability of a particular blend of semiconductors in the construction of a solar photovoltaic cell, nor even necessarily calculate how many such cells would be needed to replace a conventional generating plant.

Furthermore, as science itself becomes more complex, and as the arguments it presents become more convoluted (a subject to which I will return in the next chapter), the more irrelevant the traditional methods of science education become. There is nothing in the standard "form a hypothesis, test it, come to a conclusion" rubric of teaching the scientific method that will remotely prepare a student to deal with the latest report from the Intergovernmental Panel on Climate Change. The concept of scientific literacy, then, not only represents a new way of approaching science education for the general public, but the only way that will prepare our students for the world in which they will find themselves.

ALTERNATE GOALS

Morris Shamos and *The Myth of Scientific Literacy*

In this 1995 book the late New York University physicist Morris Shamos provided one of the first modern discussions of scientific literacy. Because of this, we can regard his work as a paradigm of the conventional wisdom in this field.

Shamos distinguished between what he called cultural, functional, and true scientific literacy. The first of these he attributes to E. D. Hirsch and the notions of cultural literacy I discussed in Chapter 2, the second to Jon Miller (see Chapter 6), while the third he reserves modestly for himself. Unfortunately, as I shall argue below, he seriously misunderstands both Hirsch and Miller, so his first two distinctions will not be of much use to us. His definition of *true scientific literacy,* however, displays an attitude seen too often among scientists (especially among my fellow physicists) and will provide a useful jumping-off point for my discussion of goals.

The basic problem with the way Shamos and others approach science education is that they seem to think that someone is "truly" scientifically literate only if he or she can come to independent conclusions about scientific issues using the same kind of reasoning a professional scientist would use. Merely having enough background to read a newspaper article isn't enough for them. In this, they echo John Dewey's "scientific habits of mind" without recognizing, as Dewey implicitly did, that such a goal is appropriate only for an educational elite. Since this is an attitude that comes up over and over again in the debate about science education, let me take some time out right here to discuss it.

Nobel laureate Sir Peter Medawar wrote the following in his 1982 book, *Pluto's Republic:*

> It is, however, an understood thing that scientists of a particularly elevated kind—theoretical physicists, for instance—may from time to time express quietly authoritative opinions on the conduct of scientific inquiry, while the rest of us listen in respectful silence. [1]

What Medawar is lampooning in this passage is a phenomenon I refer to as the "arrogance of physicists." (For the record, I trained as a theoretical particle physicist and worked in that research community until I was well past my promotion to full professor, so I have the right to comment on this phenomenon.) I think the problem is that traditional physics is a science steeped in reductionism, in which the greatest virtue is the ability to pare away complexities and come down to the essential simplicity of nature. This is an admirable quest, one that has produced one great insight after the other over the centuries. It tends, however, to produce a state of mind in which the special kinds of abilities, especially mathematical abilities, needed to carry it out are highly valued, while other kinds of abilities tend to be given short shrift.

In fact, we can use a person's attitude toward mathematics as a kind of diagnostic of his or her attitude on scientific literacy. Here is what Shamos said on the subject:

> It is unrealistic to believe that one can appreciate the broad reach of science without seeing firsthand the role played in it by mathematical reasoning. [2]

This attitude—"I had to learn a lot of difficult math to master my trade, and by God you'd better learn it too"—reminds me of the attitude of the medical community toward the ridiculous practice of having young interns work 36 hours at a stretch. (My daughter is going through this rite of passage as I write this, so I have a ringside seat to the process.) The difference, of course, is that no MD would argue that every citizen should go through something like medical school just because they want to read the health section of the newspaper, while I have heard physicists argue that everyone who wants to read about science should study calculus at the university level.

This attitude tends to have unfortunate consequences in science education because it plays into the idea, discussed in Chapter 9, that only the real thing will do in the classroom, and that only students on their way to careers in science are worth teaching. In particular, if you define the goal of science education as the production of individuals who can reproduce the kind of reasoning that scientists use to come to independent conclusions on issues, you will surely wind up with a counsel of despair, as Shamos does. In his case, concluding that the average person will never be able to understand science well enough to make important decisions, he comes out in favor of something called a Science Court, a thoroughly unworkable, not to mention antidemocratic, suggestion. In any case, as I pointed out in Chapter 2, it makes no more sense to deny students the benefits of science because they don't know mathematics than it does to deny them *War and Peace* because they can't read Russian.

I have to say that I have always regarded the idea that you should only teach those who can come up to professional standards as kind of a cop-out, a way of avoiding the real problems confronting us as educators. Instead of setting impossibly high standards and then giving up, maybe we should be thinking about how far we can go with the students we actually teach, a point to which I shall return below. In my darkest moments, I sometimes think about how easy my life would be if I could teach only the best students in my classes, label the rest as hopeless, and consign them to some sort of educational limbo.

Having made these points, let me turn to Shamos's misunderstanding of Hirsch and Miller. Since I was personally involved in both of these programs, you can think of the next couple of paragraphs as my rising to a point of personal privilege.

Shamos makes the common mistake of thinking that cultural literacy consists of nothing more than a list of terms to be memorized, a misconception already dealt with in Chapter 2. Reading his dismissive description of the concept, I can only conclude that he didn't read the books carefully or have serious discussions with any people in the field. (This conclusion is bolstered by the fact that he committed the blunder of referring to Hirsch as "Edward," a name he never uses.) At the risk of repeating myself, I have to point out that cultural literacy actually represents a potpourri of words, concepts, connections, images, and ideas, and is about as far from a dry list of terms to be memorized as it is possible to be. I challenge anyone to go through the 138 pages of science entries in *The Dictionary of Cultural Literacy* and argue that someone who has mastered that material is not scientifically literate.

His misunderstanding of Jon Miller is similar. He complained that Miller doesn't really define what he means by scientific literacy, but as I showed in Chapter 6, there is a very clear operational definition given in terms of responses to questions in Miller's standard survey. Shamos also claimed that Miller espouses an either–or definition of scientific literacy, when in fact, there are a graded series of classifications ranging from "well-informed" to "moderately well-informed" and on down. In addition, in his scholarly papers Miller publishes the complete results of his surveys in sufficient detail that anyone who wanted to go beyond his classification scheme could easily do so.

Physics First

The United States is extremely fortunate because every once in a while a scientist who has won the Nobel Prize decides to use the attendant status to improve the educational system. As you might expect of people at this level of achievement, these scientists have characteristically unique viewpoints on science education, and reviewing their work can give us a good idea of what happens when high-level scientists turn their attention to this area.

The first person to do this in modern times was Leon Lederman, who was at that time the director of the Fermi National Accelerator Laboratory outside of Chicago as well as a member of the University of Chicago faculty. Lederman is a wiry, jovial man, with a nimbus of white hair and a perpetual twinkle in his eye. Despite his stature in the scientific community, he comes across as a street-smart, wise-cracking New Yorker, which I suppose he was in his youth. After getting the Nobel Prize in 1988 for his work in experimental particle physics, he decided to see what he could do for science education. The Chicago public school system was in dire straits, as schools in many urban areas often are. "If it wasn't the worst system in the country," he commented wryly, "it was close enough that it didn't matter."

His method was simple, but direct. "The nice thing about a Nobel Prize," he once told a group of us over late-night drinks, "is that it will get you into anyone's office—once!" Lederman used that entrée to convince a number of large Chicago-based corporations, as well as various nonprofit foundations and governmental agencies, to set up an inservice program for Chicago science teachers. For reasons I never understood, it was based at the Illinois Institute of Technology, about 20 blocks north of his home institution at the University of Chicago. It was a program aimed at teacher improvement, funding substitutes so that teachers could come in, interact with working scientists, and, in general, improve their background in the subjects they taught. Emboldened by the success of this venture, Lederman went on to spearhead the founding of the Illinois State Science and Mathematics Academy, a residential school located near the Fermi National Accelerator Laboratory that provides a high-level science education for extremely gifted students.

In both of these endeavors, Lederman was following in the steps of the PSSC (see Chapter 8) and other efforts to improve science education, which concentrated on providing the best possible education for future scientists and technicians. If, as in the effort with the Chicago public schools, that also led to an increase in scientific literacy, so much the better, but scientific literacy was not his primary goal.

But as befits a man of his breadth and talents, Lederman has thought about public perceptions of science and the way that science is taught to everyone in the schools. His most important contribution to the latter has been the idea of a "physics first" curriculum. This is the notion that the current science curriculum, which usually starts with biology, should be turned on its head, with biology taught only after a firm basis in physics and chemistry has been laid. Given the fact that biologists are now concentrating so much on molecular processes, this approach makes sense from a scientific point of view.

I can see the attraction of this idea for older students; indeed, my own scientific literacy course more or less follows this scheme. The idea is to start with the simple (physics) and proceed to the complex (biology). The traditional approach, on the other hand, starts with the familiar (biology) and proceeds to the abstract (physics). The "physics first" approach has the advantage of following human intellectual history, in which the simplest problems (e.g., projectile motion) were solved first, and the more complex ones (e.g., cellular chemistry) much later. This is a new idea, so there's not much data on it, but if I had to guess I would say that a "biology first" approach is probably appropriate for elementary school and perhaps middle school, whereas the "physics first" is probably appropriate for high schools and institutions of higher education.

We Want Them to Think Like Us

Carl Weiman, then at the National Instutute of Standards and Technology in Boulder and now at the University of British Columbia, received the Nobel Prize in 2001 for producing the first example of a phenomenon called a *Bose-Einstein condensate*, a peculiar organization of atoms at very low temperatures. An amusing anecdote about a visit Weiman made to the main campus of the National Institutes of Standards and Technology near Washington to address a packed auditorium of his coworkers and guests tells how he was greeted with this song (sung to the tune of "Baby Face") :

Con-den-sate—
You've got the cutest little condensate.

The audience sang along, following a bouncing ball on the screen. Physics isn't always dry and formal!

Like Lederman, Weiman's main interest is in improving the formal science curriculum, although in his case this interest is focused on science courses for nonscientists. In a sense, he is following the Physics for Poets efforts discussed in Chapter 8. A lot of his work has been devoted to improving teaching methods through the adaptation of modern technology (e.g., having students in a large class responding via clicker to a question posed by a lecturer). His motivation, however, is very clear. Why is he spending so much time on teaching techniques? His answer is, "We want them to think like us."

And this, I suppose, is the bottom line for many scientists when they get involved in teaching in a major way, whether they have a Nobel Prize or not. In fact, the main difference between such scientists is on the question of whether the goal is achievable. Some, like Shamos, are skeptical and come to a gloomy conclusion. Others, like Weiman, are hopefully positive. But in the end, the goal is the same: to turn everyone, magically, into miniature versions of ourselves.

Why Should They Think Like Us?

When pressed on this point, most people running large science education programs eventually come back to a single argument. We are living, they say, in a knowledge-based economy, a world in which national survival depends on technological prowess. Our educational system has to produce workers to feed the new info- and biotechnologies of the future, or the country will go into irreversible decline.

It's hard to argue with this point of view; we clearly do need to produce a scientifically competent workforce as we move into the future. The problem is that there is, as discussed in Chapter 2, a difference between scientific competence and scientific literacy, and this difference tends to be blurred in these discussions. If your goal is to train a new generation of engineers, scientists, and technicians, then you want to teach people how to do science. But no matter how technological the economy becomes, it remains a fact that most people will never need to do science for a living. Everyone, however, will have to function as a citizen, and they will need to be scientifically literate to do so. The distinction between the needs of these two populations is important and, I think, largely eviscerates arguments for scientific literacy based on the knowledge economy.

For the fact of the matter is that it is not completely clear that thinking "like us" is much of a help in dealing with public issues. As I pointed out in Chapter 3, the amount of information a citizen needs for entry to public debate is pretty minimal—not at all like that of a scientific specialist. I also made the point that, outside of their field of research specialization, scientists are usually no more scientifically literate than the average well-informed citizen. Thus it seems to me that a "think like us" approach really does very little to promote scientific literacy.

The Seven Intelligences

But there is another aspect to the "think like us" point of view that I find difficult. In my darker moments, I sometimes wonder whether the rational, critical, analytical mode of thought that characterizes science isn't profoundly unnatural, in some way out of sync, with human neuronal processes. I have encountered students in my life—highly intelligent individuals who were accomplished in many areas—who have extreme difficulty thinking "like us." This has led me to suspect that there might be unanticipated difficulties in trying to turn everyone into a scientist. I know that I am deficient in certain kinds of mental skills (e.g., an appreciation of and sensitivity to colors), and I believe that virtually no amount of education would overcome those deficiencies. This is why I ask my wife to pick out my clothes when I make public appearances. I imagine that the scientific mode of thought is as alien to others as colors are to me. I find that keeping my own deficiencies in mind helps me deal sympathetically with students who are struggling in science.

In 1983 my friend Howard Gardner of Harvard University published his theory of multiple intelligences. He listed seven such intelligences initially, although he has expanded the list since then. In this list, analytical reasoning is only one kind of intelligence. For the record, his seven intelligences are:

Linguistic (word smart)
Logical-mathematical (number, reasoning smart)
Spatial (picture smart)
Bodily-kinesthetic (body smart)

Musical (music smart)
Interpersonal (people smart)
Intrapersonal (self smart)

Of these, only the second and, arguably, the third contribute to the scientific process. If you think of Gardner's categories as defining a kind of seven-dimensional IQ, then everyone has some sort of score in each category, and people who rate high in one area may well rate low in others, a fact symbolized by so many common stereotypes. To arbitrarily pick the intelligences appropriate to science and say that everyone has to excel in them makes no more sense to me than requiring that everyone write a symphony or become a performance-level dancer. The world just doesn't work that way.

All of which, I suppose, brings me back to the definition of scientific literacy given at the start of the chapter. We want average citizens to be able to deal with the scientific aspect of public issues with the same level of competence that they have in other areas. And this leads to the next, inevitable question: How much science is needed to accomplish this goal? What, precisely, does our model citizen have to know and be able to do to meet our goal?

The kinds of buzz words and phrases that get tossed around in this debate are things like "critical reasoning" and "coming to independent conclusions." In some sense, the burden of these phrases is that average citizens are supposed to be able to look at scientific arguments, listen to the competing experts, and, using their scientific knowledge and education, decide which side is right.

Let me be blunt—this is a totally unrealistic expectation. For one thing, as I shall argue in the next chapter, scientific issues today and in the foreseeable future are sufficiently complex that most Ph.D. scientists wouldn't be able to perform this task. For another, as I have pointed out repeatedly, Ph.D. scientists themselves are usually scientifically illiterate in all fields except their areas of specialty. Thus, short of requiring that everyone have a Ph.D. in everything, there is simply no way to achieve this particular goal.

A MODEST PROPOSAL

So what can we do? I would suggest that a program designed to bring every citizen as far along in science as he or she is capable of going could be based on two simple, self-evident propositions:

1. You have to teach the students you have, not the students you
   wish you had.
2. If you expect students to know something, you have to tell
   them what it is.

Before giving my detailed outline of what I believe the content of
a scientific literacy course ought to look like, let me discuss these two
propositions in a general way.

My sense is that a lot of the unhappiness I see in my fellow scientists with regard to teaching nonscientists comes from their failure to
honor the first proposition. The fact of the matter is that if you decide
to teach nonscientists (and not all of us have to make that decision)
you are going to get a mixed bag of people in your class. There will
be a small group—probably less than a third—who are genuinely
interested in the subject and really want to learn. There will be a larger group—between a third and a half—who are getting through a
required course as best they can, but are serious students who will
put in the time necessary to get a good grade. And then there is the
remainder. I sincerely hope that you do not have a large remainder in
your class.

But whatever the composition of the class, these are the students
you have to teach. Some will get it easily and some will struggle, but
the job of the teacher is to move each one as far along the path to
scientific literacy as possible. This often means that some worthwhile
goals have to be put on the back burner while you concentrate on the
science. You can't, for example, spend a lot of time correcting the students' English or writing skills—there just isn't enough time.

More important, you probably aren't going to be able to take on
the twin problems of scientific illiteracy and innumeracy at the same
time. Not only will many of these students have intelligences other
than those associated with mathematical skills, but they will also suffer
from moderate to severe math phobia. This means that if you want
them to engage with the science, you can't throw a lot of equations
at them. As I will argue in Chapter 12, the basic ideas of science can
easily be presented without equations—they are really all pretty simple
at base—so this is a problem only if teachers insist that "real" science
requires them.

At bottom, I think that the problem with my first proposition is that many scientists secretly yearn for a diminished world in which Gardner's seven intelligences are shrunk down to a world with only one or two—the one or two we're good at, of course. But the essence of being a good scientist is the ability to recognize the realities of the external world, in this case a world in which our students have many kinds of intelligence but share a common need for scientific literacy.

The second proposition also seems self-evident, but in fact is profoundly out of line with a major school of thought in science education. This is the school that holds that there is something called the scientific method (or, as Dewey had it, a scientific habit of mind), and that all we have to do is teach students this scientific process and they will grasp everything else about science on their own.

In fact, the point I am raising is an example of a long-standing issue in science education, which can be characterized roughly as the conflict between method and content. What I have been calling scientific literacy can be situated near the content side of this dichotomy, while the above is on the method side.

In fact, the proposition that we should concentrate on teaching something called the scientific method seems so ludicrous to me that I scarcely know where to begin discussing it. When I want to annoy my colleagues, I call it the "teach them relativity and they'll derive molecular biology by themselves on the way home" school of thought. There is, indeed, a scientific method—it's discussed in Chapter 1—but knowledge of this method is only a very small first step on the road to scientific literacy. In fact, I would make two arguments against the method school of thought.

First, if I applied this argument to any other field of study, its failings would be transparent. If I argued that there was something called a language method, so that studying one language (e.g., French) would give easy access to another (e.g., Czech or Urdu), we would all recognize that this argument wouldn't work. If you want to read Czech, you don't study French; you study Czech. In the same way, I would argue, if you want to discuss stem cells, you don't study climate models; you study molecular and developmental biology.

Second, as I shall argue in the next chapter, the advent of digital computers is producing a massive change in the way that science is being done. The kinds of simple experiments you can do in a school or university lab class—the kind that are supposed to teach you the scientific method—no longer have much relevance as far as many real problems in science are concerned. In the future, we can, in fact, expect the scientific method to become increasingly irrelevant to public discussions. A concentration on method rather than the actual content of scientific literacy, then, is likely to produce students ready to cope with Galileo in a world dominated by Craig Venter, the man who first sequenced the human genome.

But that is a topic for another chapter.

CHAPTER 11

Training for Galileo in the World of Craig Venter

C raig Venter is the man who is as responsible as anyone for the completion of the sequencing of the human genome, ahead of schedule and under budget. As such, he is a perfect symbol for the new kind of science that has been spawned by the advent of the digital computer, an event that has, quite literally, produced science that is fundamentally different from what has gone before. The central issue we will face in this chapter is this: How will we have to modify our educational system in order to take account of this new type of science?

The year 2000 was an epic year for the sciences. Once the worldwide New Year's Eve party was over (wasn't that terrific?), a monumental multiyear science project was completed with appropriate fanfare and a White House press conference. This was a project aimed at completing what was called the "first assembly" of the human genome: a step-by-step catalogue of all 3 billion "letters" that, taken together, constitute the human genetic endowment. If you think of the DNA molecule as a kind of twisted ladder, then the rungs of that ladder are made of two linked molecules called *bases*. There are four bases, designated by the letters A, C, G, and T, which form a kind of four-letter alphabet in which the language of all life is written. The complete list of these bases, or *genome*, is the sum total of all information passed from one generation of any organism to its offspring, and it contains all the information needed to run that organism's chemistry. The process of working out this code is called *sequencing* the genome.

When the idea of sequencing the entire human genome was first proposed in the 1980s, there was a fair amount of opposition to it from traditional biologists. Up to then, biology had been run in a mode called *small science*—a typical research operation was a professor and a couple of students pursuing their research in a basement laboratory somewhere. The corporate nature of the genome project, with billions of dollars pouring in, made some biologists nervous. In addition, the techniques available at the time tended to be complex and labor intensive, so the project seemed to promise a lifetime of drudgery for those caught up in it. As one grad student told me, "I don't want my life's work to be sequencing from base pair 100,000 to base pair 200,000 on chromosome 12." (He has since become a prominent paleontologist.)

These fears turned out to be unfounded, because a massive worldwide effort was soon under way to find ways of automating the sequencing of DNA. It was at this point that Craig Venter came on the scene. A dynamic, unconventional individual, Venter is the type of person who elicits strong reactions from people around him. Richard Preston pointed this out in a profile of him in *The New Yorker*, which started off with the anonymous quote "Craig Venter is an asshole." Venter began working on the genome project under the auspices of the National Institutes of Health, but quickly became impatient with the plodding pace of the federal program. He left his job and founded Celera Genomics, a private corporation dedicated to finding fast, efficient ways of sequencing genomes. Along the way he developed a technique for carrying out the sequencing operation that brought the project in years before its projected completion date and cost several billion dollars less than had been originally thought. In the process, he also provided an illustration of the major sea change associated with computers that I mentioned above.

His technique, known as the *shotgun*, involves breaking up stretches of many copies of DNA into short pieces, feeding these pieces through parallel automatic sequencing machines, and then having a computer match up the pieces and "assemble" the entire original genome. It would be like reading a book by first ripping up many copies, handing each scrap of paper to a separate reader, and then having computers reassemble the original text from the data supplied by all of those readers. In Venter's shop, the computers were as important as the actual biological sequencing.

The story of the Human Genome Project, then, illustrates the fact that in every field of science, computers are coming to play critical roles in forefront research, with the kind of impact produced dependent on the nature of the field itself. This fact is changing the way that science is done and, consequently, changing the kind of science that will be presenting itself in public issues. It is, therefore, worth taking some time to think about what computers are and how they function in the scientific world.

THE ROLE OF COMPUTERS IN SCIENCE

A computer is a tool—a marvelous, multifaceted, complex tool, but a tool nonetheless. Like all tools, it does some things very well—better than humans—and some things very poorly. This is, after all, the nature of tools. Your car can travel faster than you can, but it can't balance your checkbook. For people who are worried about computers taking over, I offer philosopher John Searle's reassuring remark: "No one worries about shoes taking over the world, so why worry about computers?" [1]

From the point of view of science, there are two capabilities of computers that are important: data management and computation. Every field of science is exploiting these two capabilities in different proportions and, in the process, transforming itself into something new.

Take paleontology as an example. Thirty years ago this was a field devoted almost entirely to the detailed study of individual fossils, each painstakingly and lovingly measured and catalogued. The image of the paleontologist was either of the rugged scientist out in the field collecting fossils or the bearded savant in the back room of a museum analyzing them. One by one, the fossils were entered into the *fossil record*, the compendium of all scholarly articles on the subject.

The problem was that no one was able to look at the record as a whole because there were just too many details. It was a classic case of not being able to see the forest for the trees. Then in the late 1970s Jack Sepkoski at the University of Chicago began to change all that. Delving into the literature, he began to assemble a computerized record of all reported fossils. He used to joke about this in his seminars, showing a picture of the university library and claiming that it was his field area. But as his database grew, the power of the computer began to manifest itself. No human brain could possibly remember the thousands of entries in the fossil record, but the computer can. Once the memory capacity of the human brain was augmented, patterns began to emerge from the data. One of the first important results, for example, was a paper Sepkoski wrote with his colleague Dave Raup in which they showed that the pattern of extinctions in the Earth's history follows a repeating pattern, with a large extinction event (like the one that did in the dinosaurs 65 million years ago) coming every 26 million years.

Once the power of this use of computer memory was clear, its use spread like wildfire in the community. Today, a paleontologist is expected to be as skillful with a database as with a geologist's hammer, as much at ease at the keyboard as in the field. Large-scale questions—Who survives extinction events? Who succumbs? Do organisms really evolve to greater size over time?—have been addressed and answered. You could no more do paleontology these days without a computer database than you could do it without fossils.

The computational ability of the machines has played a similarly transforming role in other fields. The basic point is this: Much of science consists of building models of the external world, models whose purpose is to mimic that world and predict its behavior. Typically, these models are expressed in terms of mathematical equations. For example, when Isaac Newton produced his model of the clockwork universe, the key element in his theory was his law of universal gravitation, which is written

$$F = GMm/r^2$$

where F is the force between two objects, M and m their respective masses, r the distance between them, and G a universal number known as the gravitational constant.

Once an equation like this is written down, the next task is to solve it in a specific context, for example, to work out the orbit of the moon or the tides in the Earth's oceans. And that's where the problems start, because as soon as you get past a minimal level of complexity, traditional pencil-and-paper methods of solution just break down.

Consider the motion of objects in the solar system as an example. At any moment, there are many gravitational pulls acting on a planet like the Earth. The sun exerts a big one, of course, but so do all of the other planets and moons. If you imagine freezing the moving planets for a moment—hitting the pause button, if you will—you can calculate the *F* in Newton's equation just by adding up all of these individual forces from the known masses of the other bodies and their distance from the Earth. With this knowledge, you could calculate how all of those gravitational forces will make the Earth move. You could, if you wanted to, do a similar calculation for every other object in the solar system.

So far, so good, but now things start to get complicated. If we take our finger off the pause button and let the planets and moons start to move, they will begin by reacting to the gravitational forces we just calculated. A few seconds later, however, *everything will have moved*. This means that all of the forces will have changed, and the *F* we calculated during the pause will no longer be the actual force acting on the Earth. To deal with this, we have to hit pause again, re-calculate the forces, find the new directions of motion, and let things go forward again. By stepping forward this way, alternating pause and go, we can, in fact, predict the future motion of every object in the system.

This is a complicated process, involving a lot of calculation. Traditionally, the way scientists have approached it is to try to build a simplified model of the solar system by ignoring some of the effects that we suspect are small. The first step on this process might be to consider only the Earth and the sun, for example. In this case, we can use calculus to work out the planet's orbit with pencil and paper (it's a simple ellipse). It was, in fact, this type of problem that Newton had in mind when he developed his version of the calculus. If, however, we depart from simplicity and add in something else—the moon or Jupiter, for example—we can no longer get a pencil-and-paper solution, but find ourselves back in the kind of complexity described above.

In the late nineteenth and early twentieth centuries, astronomical observatories routinely employed a large staff whose sole function was to carry out these kinds of laborious numerical calculations using tables of logarithms. Interestingly enough, these people were called *computers*. I can remember seeing an obituary in a science journal years ago, when the last human computer died. But there are limits to the complexity of the calculations that can be done this way, since human computation is relatively slow.

The advent of modern computing machines, of course, changed all that. Not only can a computer do in seconds a calculation that could take humans a month to complete, but the computer can handle a lot more complexity. It really makes very little difference to the computer whether you stop at adding the moon and Jupiter or throw in all the planets, their moons, and a few dozen asteroids. It's just takes a few extra milliseconds to add in their forces when you calculate F.

Since the 1980s, this ability to find numerical solutions to complex equations has revolutionized many fields of science. In the 1970s I got involved in the field called *fluid mechanics*, the branch of physics that deals with problems like the flow of liquids or gases. At that time, the frontier of computation with the computers we had was the calculation of the flow of air over a thick airplane wing moving near the speed of sound. In those days, new airplane designs had to be tested by putting models into massive—and expensive—wind tunnels. By the end of the 1980s, computers had improved to the point where an entire airplane could be designed by using the kind of numerical techniques outlined above. The stresses and strains on every part of the plane in any kind of situation could be worked out, and this told the engineers how much strength was needed in each part of the structure. The Boeing 777 was the first commercial aircraft designed completely by computer. In the design of this aircraft, all plans were stored in computers, and many were tested by having engineers, and even airline customers, take virtual walk-throughs of the plane before any pieces of metal were ordered or cut.

In other words, in this field we don't experiment on physical models anymore, or, more precisely, the models on which we experiment are inside a computer, not out in the real world. The term we use for this kind of construction of an aircraft is *computer modeling*. In effect, our new computational ability allows us to build models of the real world inside our machines. Over the past couple of decades, as Moore's law (the law that says that the speed of computers will double every 18 months) has made computation ever faster and cheaper, we have been able to model more and more complex systems in this way.

In fact, today you can find computer models of varying levels of complexity (and believability) in areas as widely separated as predictions of future climate (a subject to which we'll return in a moment), nerve signals in the brain, and behavior of the stock market. In most of these fields, the main barrier to progress is the speed and capacity of computers. As Moore's law has its inevitable effect, however, scientists are able to include more and more detail in their models, which means that the models are starting to behave more and more like the real world. (Here's an interesting way to think about the effect of Moore's law: a calculation that can be done today only on the world's best supercomputer will in 15 years be the sort of thing that you will be able to do on your personal laptop or work station.)

This means that the sheer growth and availability of computational power not only is going to continue, but it will have important effects on the entire scientific endeavor. There are already two new fields of science—chaos and complexity theories—whose origin depends entirely on the availability of computers. What will the traditional branches of science look like by the time today's elementary school students take on their role as citizens? It is to that question that I now turn.

THEORY, EXPERIMENT, AND THE THIRD WAY

Traditionally, scientists have been split into two camps, depending on the sort of work they did. On one side were those who worked in laboratories or observatories, discovering the nature of the world we live in. On the other side were scientists who tried to interpret these results and produce models (usually mathematical) representing that world. These two camps are referred to as *experiment* and *theory*, respectively. Typically, scientists made a choice between these two grand approaches sometime about the second year of graduate school and then spent the rest of their careers following the consequences of that choice.

Because of this, we thought of scientific progress as a kind of dialectic waltz through history, with theorists making predictions and experimentalists then either verifying or negating them in the laboratory. The advent of computer modeling has challenged this neat categorization scheme. A computer model isn't really a theory, but it's not really an experiment either. It's something else, and many scientists are starting to think of computer modeling—or, more formally, *computational science*—as a third way of doing work in our craft.

A computer model starts with theory—in the case of the airplane, for example, with the known laws that govern the behavior of compressible gasses like air. It puts in a lot of other things that may come from theory or may come from experiment. For example, there may not be a theory that describes the friction between a certain kind of metal surface and air at a particular temperature or pressure. What may happen then is that the results of experimental measurements of this quantity are fed into the model instead of theoretical equations. In the end, the inputs into the model represent our best knowledge of all of the factors that might be important in the design of the aircraft. At this point the computer takes over and crunches its way through the numbers, finally spitting out whatever quantities we have asked it to compute.

In general, the kinds of computations done in this way are much too complex to be checked, step by step, by a human being. It would just take too long. Thus the actual calculations are usually something of a black box—you feed numbers and equations in here and the answers come out there. And this gets us to the basic problem encountered in this new area of science: How can you trust the results of a model when you don't really know how the results were derived?

In the same way, because of the complexity of the calculation, it is extremely difficult to tell whether some approximation made in a part of the program threw the results off or not. For example, if you represented the roughness of the airplane surface as a series of little spherical bumps (think of a series of ping-pong balls cut in half and glued to the wing) instead of a more realistic but more complex shape for the wing's surface, are the results of the calculation truly representative of what happens on a real airplane wing? The basic question comes down to this: How do you know that the world you've created inside of your computer model is actually the same as the world in which we live? This is called the problem of *validation* of the model, and it is a crucial, though often underappreciated, part of the third way of doing science.

There are many ways of going about validating a model. If we are making a model of a real system (e.g., an airplane, or a forest ecosystem), we can see if the behavior of the system in the computer matches the behavior of the system in the real world. This sounds simple, but consider this: Any model of moderate complexity will make many, many predictions. The biggest model of forest ecosystems I've ever seen is run by the Department of Agriculture, and it will make predictions about things like the types of trees that will flourish in a given area, the number of board feet of lumber a particular stand will produce, and so on. What usually happens is that you find that some of these predictions are borne out by observation while others are not. And that, of course, raises a serious question: What percentage of observations have to match the predictions before we say that the model is a good representation of reality?

This is an important issue, because it requires a kind of judgment that goes well beyond the prediction-experiment-confirmation or denial method used in the standard teaching of the scientific method. If the model is to be used to shape public policy, as it could well be in the case of the forest ecosystem models mentioned above or the climate models I'll discuss below, then citizens will have to have at least some sense of how to make judgments like this.

The complexity of the calculations in many models raises another frustrating problem. Even if the model results match what we see in the real world, how can you be sure that you haven't gotten the right answer for the wrong reason? How do you know, for example, that you didn't make two mistakes that happened to cancel each other out in this calculation, but might not do so when we use the model as a basis for future policy? This is a difficult question.

Another way to test the results of a model is to change some of the inputs and see whether the final results depend on that change. In the example of airflow over a rough wing given above, for example, you could change the size and spacing of the ping-pong balls and see if any of the predicted flight characteristics change. If they don't, then you can say that even if your model of roughness isn't exactly right, any discrepancies that exist won't make any difference to your model. Scientists call this process *exploring parameter space.*

The problem with this approach is that it often takes a lot of computer time to run the models, and there are often hundreds or even thousands of parameters that could be varied. This means that the amount of exploration you can actually do is often quite limited compared to the total possible number of variations. It is not unusual for the kind of climate models I'll discuss below to take months of computer time to predict the climate 100 years from now. So again we are confronted with a judgment call: How much exploration of parameter space do we have to do before we are confident that our ignorance of the details of a particular process won't affect the final outcome? How can we be sure that the next computer run (the one we don't do) wouldn't diverge wildly from what we have already calculated?

The central issue we have to face from the educational point of view, then, is how to go about familiarizing students with this new kind of science, with all its complexity, so that they can make the kinds of judgments that will be required of them later in life. As is suggested by the title of this chapter, I think that while the standard kind of laboratory experiments students encounter—rolling balls down inclined planes, for example—help them understand the relatively simple world of Galileo, they won't get students very far in the brave new world of computational science. At the very least, they should know that this new world exists, and that simple yes or no answers to scientific questions are going to be harder to come by in the future than they were in the past.

CLIMATE MODELS: A CASE IN POINT

As of this writing, global warming is once again claiming public attention. Politically speaking, this issue constitutes a "perfect storm," involving as it does complex science, unpopular personal and political choices, and the promise of immediate pain for a distant and uncertain payoff. There is nothing simple about it, and consequently it can serve as a good example of the kind of discussion for which we have to prepare our students in the twenty-first century.

For the record, we know that the planet has been warming since the end of the Little Ice Age in the mid-nineteenth century. The real question is how much of this warming is natural, and how much is due to human activity.

The basic physics of global warming is not controversial. Human beings, by burning fossil fuels, are adding carbon dioxide to the Earth's atmosphere. The carbon dioxide acts as a so-called greenhouse gas, absorbing infrared radiation given off by the planetary surface and, in effect, providing a kind of blanket that, other things being equal, can be expected to raise the Earth's temperature. The question: What changes in temperature and climate will result from a given addition of carbon dioxide?

There is only one Earth, so we can't do experiments in which we add different amounts of carbon dioxide to the atmosphere and wait to see what happens. This means that the simple predict-observe-confirm scenario discussed in Chapter 1 isn't going to help us much in the global warming debate. Only computational science can be used to answer the question of what effect human activities will have on the planet, and that, in turn, means that we have to construct a computer model that includes everything that could affect the climate. Such models go by the name *global circulation models,* or GCMs. I want to spend a little time describing these models so that the complexity of the calculations is clear.

There are many difficulties involved in producing a GCM. For one thing, there is a fair amount of carbon dioxide in the Earth's atmosphere regardless of human activities. In fact, if it weren't for this natural greenhouse, the oceans would have frozen over billions of years ago and the average temperature of the planet would be about 18 degrees below zero Celsius or a few degrees below zero Fahrenheit. The expected warming from human activities is a matter of a few degrees C, so the model has to be able to calculate what is, in effect, a small correction to a big natural effect. This is always a difficult job. (It is customary to report expected warming in degrees Celsius [C]. For the rough estimates I'll be using here, just double the change in Celsius degrees to get a close equivalent in degrees Fahrenheit.)

It is easy to describe the construction of a GCM, considerably less easy to carry the task out. You begin by breaking the atmosphere and ocean up into discrete boxes, with the size of the boxes depending on the power of your computer (a point to which I'll return later). We hit the pause button and apply the known laws of physics to the fluid in each box, calculating things like temperature changes, energy flows, condensation or evaporation of water, movement of specific gases, and so on. Once this has been done, we can step the model forward, hit the pause again, and repeat the process, taking account (for example) of the movement of fluids or energy from one box to the next. In a typical computer program, such as the kind used to generate your evening weather forecast, the boxes will be some tens of kilometers (km) on a side, with 11 boxes stacked from the surface to the top of the atmosphere, and the program will step forward 20 minutes between pauses. Typically, this process produces some tens of millions of boxes for the computer to deal with. These numbers represent the limit set by the computational speed of the machines coupled with the fact that you have to be able to produce tomorrow's forecast before tomorrow actually arrives. If you want to calculate the climate a hundred years from now, you have to simplify the calculation by expanding the size of the boxes to more than a hundred kilometers on a side and run the computer for months instead of hours.

There are literally hundreds of quantities that have to be fed into a GCM before the run starts. For example, we know that ice reflects incoming sunlight while water absorbs it. If the temperature goes up in the world inside of your computer, you have to estimate the amount of ice that is melting between each pause and adjust the heating due to incoming sunlight accordingly. Quantities such as the abundance of vegetation, cloud cover, and aerosol particles in the upper atmosphere, along with myriad other factors, have to be put in before the GCM can make a prediction. As outlined above, you have to worry about whether any of them could be wrong and, if they are, what effect that has on the outcome of the computer run. Given the time each run takes, it should be obvious that the task of exploring parameter space for these models is prodigious, to say the least. This is one problem we have to think about in dealing with their predictions. (Another problem is that a mesh several hundred kilometers on a side cannot do a good job of representing major storms like hurricanes.)

There is yet another problem. We know that clouds play a major, though complex, role in determining climate; some clouds tend to trap heat on the surface, while other ones tend to reflect sunlight. We also know that clouds seldom occur in chunks several hundred kilometers on a side, which means that the fact that we have to use boxes this size makes it hard for us to take account of cloud effects accurately. This is a limitation imposed on us by the current state of most of our computers. When the giant supercomputer called the Earth Simulator came online in Yokahama in 2002, this limit began to be lifted. That machine, built specifically to model the Earth's weather and climate, is fast enough to reduce the size of the boxes to about 10 kilometers on a side, with a serious possibility of getting down to 1 kilometer. Indeed, so great is the detail in these calculations that watching the machine's output is like watching a film taken from a weather satellite. In any event, this aspect of the problem of clouds will eventually succumb to Moore's law as computers get faster.

But once we get the size of the boxes down to the size of clouds, we encounter another, more fundamental difficulty. The fact of the matter is that we just don't know enough about how clouds form and operate to predict what will happen when, for example, air of certain humidity cools a certain number of degrees. To the extent that clouds are important to predictions of future climates, then, this ignorance will introduce significant uncertainties into the predictions of the GCMs.

Because of these and other kinds of uncertainties, different groups of scientists come up with different scenarios for the Earth's future climate. It is customary to talk about the amount of warming that would accompany a doubling of carbon dioxide levels above the pre-industrial levels as a kind of benchmark in this discussion (current levels are about a third higher than preindustrial levels). In 2007 the Intergovernmental Panel of Climate Change, a large international body that comes as close as any to representing a consensus of scientific opinion, projected this temperature increase to be between 1.5 and 4.5 degrees C, with a best estimate of 3 degrees C. For reference, the Earth has warmed a little less than 1.5 degrees C since the end of the so-called Little Ice Age in the mid-nineteenth century, while the transition out of the last Ice Age 10,000 years ago involved a temperature swing of about 5 degrees C.

In the best of all possible worlds, we would be able to base policy decisions about greenhouse mitigation on sound cost-benefit analysis. We would be able to say things like "adding X tons of carbon dioxide to the atmosphere will cause this much temperature increase, which will have these costs, while curtailing these emissions will cost this much in additional facilities or lost economic activity." There would be many arguments about such a comparison—How much value do we assign, for example, to the loss of an obscure plant species or to undesirable changes in a forest ecosystem? The point, however, is that because of the uncertainties in the GCMs, we can't even get to that debate because we can't say, with any meaningful degree of certainty, what the climate consequences of a given action will be.

This situation is, as I said, a foretaste of things to come as far as public debates are concerned. The science involved is very complex, and probably not well understood by anyone outside of a small circle of experts. The standard methods of validation of the GCMs likewise yield ambiguous results. When we ask if they give an accurate description of the real world, the answer is that they do well in some areas, not so well in others. If, for example, you ask them to "predict" the climate of the twentieth century, they cannot do so from first principles, although they can get reasonable predictions by adjusting various parameters.

EDUCATIONAL IMPLICATIONS

We can use the problem of global warming as a paradigm case of the science of the future and ask what sorts of knowledge our students will need to deal with it intelligently. Obviously, we can't expect everyone to construct their own GCM and verify the predictions made by scientists. Nor can we expect everyone to be conversant with the details of the different approximations that go into the models—most scientists (the author included) really don't know much about this. We have to think, then, about precisely what sorts of issues we want average citizens to be able to deal with.

The most important educational task, I believe, is to make sure that average citizens know what kinds of questions ought to be asked in a given situation. In the case of global warming, for example, we can approach this problem by thinking of the scientific issues as being composed of a layered set of questions, each more general than the last. The question at the bottom concerns the individual inputs into the computer model: For example, did we get the sea ice changes right? This is a purely scientific question, one probably best left to the experts. The next question involves what happens when these inputs are put into a GCM: Will the final results be sensitive to whatever uncertainties there are at the first level? How much of parameter space has been explored, and is this enough? The answers to these questions begin to tell us the amount of faith we are going to put in the GCM predictions. At the next level we face the problem of validation: Do the descriptions of the world in the computer match the world we actually live in? This is a question that will be debated publicly by scientists, and one that average citizens can follow. Finally, we have to face what is perhaps the most difficult problem: Given that the Earth's climate is always changing, how much of the current warming is actually due to human activity? It is only after we get through all of this that we can get to the true bottom line: What are we going to do (or not do) about global warming? No matter how complex the science behind debates in the future, the outstanding questions will be layered in this way.

What background knowledge do average citizens need to deal with these layered questions for themselves? First, of course, they will need to know enough about the basic science involved to understand how predictions of global warming are generated and have some sense of where the various uncertainties might come from. They should also know enough about the methods of science to understand that even though science may not be able to produce certain answers to important questions right now, it nonetheless can provide guidance for decision making and eventually produce a picture with much less uncertainty. In the end, they should be able to deal with the scientific predictions with the same sense of intelligent skepticism that might be used in dealing with other uncertain forecasts, such as those of economists or meteorologists. For the fact of the matter is that people are perfectly capable of dealing with uncertainty in making public decisions—they do it all the time in nonscientific areas. I see no reason why the average person cannot approach complex scientific issues in the same way, taking expert opinion into account while at the same time preserving a certain healthy skepticism.

But what about the issue on the front page *today*? What, in other words, about the need to make an immediate decision in the face of uncertainty? I argue that once students have acquired the basic framework I have called scientific literacy, it is relatively easy for them to understand the nature of computer modeling. It is also relatively easy to teach them to judge the relative authority of the various participants in any scientific debate. The existence of uncertainty, in other words, may make it harder to come to a decision but it needn't stop us from making provisional judgments. Once average citizens have this basic background, it's time for scientists to adopt a hands-off attitude and let people make up their own minds. That, after all, is what democracy is all about.

CHAPTER 12

The Great Ideas Approach to Scientific Literacy

S everal months before this writing my wife and I moved into a new home in a new subdivision near George Mason University. As a former construction worker, I have enjoyed walking around watching the new homes going up, chatting with the builders, and, in general, playing the role of sidewalk superintendent to the hilt. You don't have to understand construction, however, to appreciate one of the central facts of building: No matter how fancy and complex a building turns out to be in the end, every project starts with laying a firm foundation. It may be the concrete walls of an excavated basement, steel beams driven down to bedrock, or a simple cement slab, but it has to be there before you can go forward.

Scientific literacy, the background knowledge about science that people need to function as citizens and appreciate the world around them, plays the same role in science education that those concrete foundations do for the houses going up around me. Basic knowledge of science has to be in place before you can start to talk about more advanced concepts, just as the foundation has to be in place before you can build the rest of the house. To my mind, people who argue that we should be teaching students the scientific method or giving them a scientific habit of mind are, in effect, trying to build the house before the foundation is laid.

Before people can start to think critically about any subject involving science, they have to know something about what that science is. There is no point, for example, in trying to teach students to think critically about global warming if they don't know the basics of planetary energy balance. In the end, you cannot think critically about "nothing"—the concepts you manipulate have to be in your mental arsenal before you start manipulating them. This notion is the basic logic behind cultural literacy, and was discussed at length in Chapter 1.

In this chapter I want to talk about a natural way of building this foundation for all of our students and then move on to speculate about what building the rest of the house might look like as far as science education is concerned.

THE GREAT IDEAS APPROACH TO SCIENTIFIC LITERACY

When I think about what I like to call the scientific worldview, I often use a simple spiderweb as an analogy. Around the edges of the web are all the phenomena of the physical universe—beaches and baseballs, stars and spiders, tangerines and towers. Start at any point of the web and begin asking questions: What is this thing? How does it work? How is it related to the stuff around it? As you do this, you work your way into the body of the web, discovering unexpected connections along the way. You find out, for example, that a lightning bolt and the force that allows decorative magnets to keep notes pinned to your refrigerator are related, and that both are related to the electricity that runs your house and the light that allows you to see. Eventually, when you have worked your way to the very center of the web, you find a small number of general principles that, taken together, explain the operation of the entire universe. Call them Universal Principles, Great Ideas, or what you will, they are the core, the skeleton, of the way scientists look at the world. They are what give our view of the universe its shape and form.

This natural hierarchy in the organization of the sciences suggests an approach to science education based on the Great Ideas. These ideas (which I will list explicitly below) form a kind of superstructure to the edifice of science. They are the framework on which everything else can be hung, as walls and windows can be hung from the steel girders of a skyscraper. It is my contention that someone who has this framework in place—who has, in other words, mastered the Great Ideas—will be scientifically literate.

The fact that the Great Ideas can be thought of as a kind of intellectual skeleton allows an enormous flexibility in the details of how material is presented, which is always a useful property for an educational scheme. When I want to illustrate this point, I talk about an experience I had when I was working on the first version of *The Sciences: An Integrated Approach* (see Chapter 8). A number of instructors around the country had agreed to teach from our first draft, and my coauthor Robert Hazen and I traveled around to visit their classrooms. Thus I found myself in a class at DePaul University in Chicago, listening to Lynn Narasiman, a professor of mathematics, giving a lesson on pseudoscience to a group of about 20 future teachers.

In the book we had suggested a simple experiment (described above in Chapter 1) to test astrology in which students are given yesterday's horoscopes and asked to pick the one that would have been the best advice to have had the day before. We remarked that you would expect one student in twelve to get the "right" answer just "by chance." Narasiman had the students go through the exercise, got the expected result, but then turned the class into a wonderful hour-long exercise on the concept of chance—a discussion that introduced the students to the basic ideas of probability theory.

The point of this example is that it would never occur to most of us (certainly not to me) to use astrology in this way. But whether I would think of this particular use of the Great Ideas doesn't matter, because the system allows each instructor to bring his or her specific expertise and interests to the subject at hand. Thus each will hang his or her own ideas on the framework provided by the Great Ideas.

Finally, the Great Ideas share an important quality with other notions connected to cultural literacy. None of us knows what the subjects of public debate will be 20 years from now; certainly no one would have foreseen the current discussions of genetic engineering or stem cells 20 years ago. But whatever the subjects of future debate turn out to be, I can guarantee that they will be connected to the framework of the Great Ideas.

I will list and discuss these Great Ideas, in an order that I have found works well from a pedagogical point of view.

The universe is regular and predictable.

In a sense, this is a necessary condition for doing science. If you let go of this book and it fell up as often as it fell down, you really couldn't talk about applying the scientific method at all. The method depends on the notion that repeated experiments or observations will give the same results; this is why we can find regularities in nature and construct theories. Newton's laws of motion and the law of universal gravitation are excellent exemplars of what happens when this Great Idea is allowed to play itself out in the real world.

There has been some discussion in recent years to the effect that the advent of chaos theory invalidates this particular Great Idea. This argument, I think, arises because people do not pay attention to the fine points of the nature of chaotic systems. Let me make a short digression to clear up this point.

A system is chaotic if starting it off from two separate but close initial situations produces two widely different final outcomes. For example, two chips of wood placed next to each other on the upstream side of a rapids will, in general, come out far apart from each other on the downstream side. To make a prediction about the behavior of a chaotic system, then, you need to have extremely accurate—in principle, infinitely accurate—information about the system's starting point. With this information, you can, indeed, calculate its future behavior precisely. In general, however, there will be some uncertainty in those initial measurements, and this, in turn, means that you won't be able to make such a calculation. In other words, although chaotic systems are predictable *in principle*, they may very well be unpredictable *in practice*.

This aspect of chaotic systems doesn't mean that we can't know anything about them. The Earth's atmosphere is probably chaotic, for example, but we can still use computer models to make good enough predictions of the short-term weather.

The energy of a closed system is conserved.

Heat will not flow spontaneously from a cold to a hot body.

Known, respectively, as the first and second laws of thermodynamics, these two statements describe the nature and movement of energy. The first law, for example, in its alternate form (Energy cannot be created or destroyed) provides the basis for the science behind global warming. It tells us that the energy captured by greenhouse gases doesn't just disappear. It has to go somewhere—in this case, to warming the planet. The second statement deals with fact that the universe (or any closed system, for that matter), as a whole, becomes more disordered over time. It has some unexpected everyday consequences as well. It tells us, for example, that when we burn coal to generate electricity, fully two thirds of the energy in the coal must be dumped into the environment as waste heat.

Maxwell's equations govern electricity and magnetism.

These equations are named for the Scottish theoretical physicist James Clerk Maxwell (1831–1879) who first wrote them down. They tell us, for example, that there are two kinds of electrical charge, with like charges repelling each other and unlike charges attracting. They tell us that moving electrical charges (currents) produce magnetic fields—the basic operating principle of the electric motor—and that changing magnetic fields produce electrical currents—the basic operating principle of the generator. In addition, Maxwell used these equations to predict the existence of waves from radio and microwaves on up to X-rays. It was the research that led up to these equations that produced today's electrified society, and was the background of Faraday's remark to the prime minister ("Someday you will be able to tax it") quoted in Chapter 7.

Matter is made from atoms.

This Great Idea deals with the basic structure of matter in the universe. There is a long history of work, from ancient Greek speculation to the modern data-driven atomic theory, which has led to this particular insight. It is also the first of the Great Ideas that is, in a sense, still open-ended, since the quest begun by the Greek atomists has continued down through nuclear physics, particle physics, quarks, and in the future may even move on to things like strings.

Think of it this way: In the early nineteenth century, the British chemist John Dalton introduced the modern idea of the atom. In keeping with the meaning of the Greek word ("that which cannot be divided"), Dalton thought of his atoms as indivisible. If you picture Dalton's atoms as bowling balls, you won't be far off. By the early twentieth century, however, we knew that atoms had a complex internal structure, with electrons circling in orbits around heavy nuclei. By the mid-twentieth century we had discovered that hundreds of different kinds of short-lived particles existed inside the nucleus, and shortly thereafter we realized that those so-called elementary particles are themselves made up of things more elementary still—things that were dubbed *quarks*. Today, theorists think that the quarks themselves might be manifestations of vibrations of even smaller, more complex objects called *strings*. The quest to understand the basic structure of the universe—what physicist Steven Weinberg called the *Dreams of a Final Theory* in his book of that name—goes on.

The properties of materials depend on the identity, arrangement, and binding of the atoms of which they are made.

Once you know that all material objects are made from atoms, you can start talking about the formation of the infinite number of materials around us, from our own bodies to microchips to the metals and plastics in your car. We have to understand how atoms come together and form bonds in chemical reactions to understand the kinds of properties—mechanical, electrical, magnetic—the resulting materials will have.

A particularly important aspect of this subject is the behavior of *semiconductors*, those silicon-like materials on which the modern electronic industry is based. As the name implies, they are not really either conductors (like copper) nor insulators (like rubber), but something in between. It turns out that this property allows us to exercise exquisite control over electrical currents, a fact that led to the development of the transistor and, ultimately, to computers and the entire modern digital, information-based economy.

In the quantum world you cannot measure an object without changing it.

With quantum mechanics we enter the strange world of twentieth-century physics. For the first time, we encounter phenomena that do not fit the comfortable Newtonian paradigm. This is, then, an excellent place to talk about how changes occur in mature sciences. We do not, for example, discard Newtonian mechanics because it doesn't work inside the atom. Instead, we recognize that if we apply the laws of the quantum to baseballs and spaceships, the unfamiliar new laws reduce to the familiar old ones. This means that mature sciences grow by accretion, like rings in a tree, adding new ideas without replacing the old. Science no longer changes by replacement—an important lesson in philosophy.

Having said this, I also have to say that quantum mechanics and relativity (see below), although intellectually fascinating, are two fields that deal with areas far from everyday experience, and are therefore least likely to be encountered by people in their role as citizens. Thus these subjects should be thought of as being an example of the science as culture discussion in Chapter 4, with the argument being that these areas should be a part of everyone's cultural repertoire, like medieval European history.

The laws of nature are the same in all frames of reference.

Known as the principle of relativity, this is the basis of Einstein's famous theories. The phenomena of relativity are fascinating: moving clocks slowing down, nothing moving faster than the speed of light, $E = mc^2$ and all that. Like quantum mechanics, however, relativity deals with things far removed from everyday experience.

There is a great deal of energy in the atomic nucleus.

The nucleus is made of particles, which are made of quarks. . .

The nucleus of the atom enters the domain of scientific literacy in several areas; it's involved in nuclear power (both fission and fusion), radioactivity, the use of radioactive tracers in medicine, and nuclear waste disposal, to name a few. To understand the discussion of these issues, people need to know some basic things about how a nucleus is put together and how the energy within it can be tapped. Once we get past this level, down to the elementary particles, quarks, and strings discussed above, we once again come into a domain in which many people find a high level of interest, but arguably a domain with little current application in the everyday world.

Stars live and die like everything else.

The universe began in a hot, dense state about 14 billion years ago and has been expanding ever since.

These ideas, which deal with fields that astronomers call galactic dynamics and cosmology, respectively, deal with the place of the sun (and therefore the Earth) in the cosmos. Like relativity and quantum mechanics, these subjects have little everyday import, but can be thought of as essential pieces of every educated person's cultural apparatus. In a society where so much science fiction (both good and bad) is available, and which routinely spends tens of billions of dollars each year on space exploration, they also provide some needed background knowledge for budgetary debates. In the same vein, a discussion of stellar lifetimes can get you to *supernovae* and *black holes* (always topics of high interest).

The surface of the Earth is constantly changing.

The advent of the theory of plate tectonics (our current picture of the operation of the Earth) in the 1960s is one of the best examples of how scientific change can be driven by new data. In a matter of a few years, the entire Earth-science community gave up their old "fixist" ideas and came over to the new dynamic picture of our planet. In this picture, slabs of rock 30 to 50 miles thick, called plates, make up the Earth's surface, moving around in response to the flow of heat ("boiling") in the deep interior. As a consequence, our own North American plate, which extends from the middle of the Atlantic to California, is moving farther from Europe at the rate of several inches per year. The surface features of the Earth are in a constant state of flux; nothing lasts forever.

Once we've established the impermanence of everything from mountains to oceans (a surprising concept to most people) we can also use this discussion as a takeoff point to the rest of the solar system ("Why is the Earth so different?") and extrasolar planets ("What else is out there?"). Both of these are excellent examples of the phenomenon of hanging things from the framework provided by the Great Ideas.

As a side issue, understanding the fact that the Earth is constantly changing can help students avoid one of the most egregious attitudes

I see in environmental debates. I call it the "fallacy of the snapshot." It's the idea that every aspect of the Earth has always been the way it was when we first became aware of it and any change represents some sort of disaster.

The Earth works in cycles.

I like to think of the Earth as a giant machine with all sorts of wheels turning and energy flowing through. The wheels are the great cycles on the Earth's surface. The rock cycle, operating over hundreds of millions of years as new molten rock comes to the surface, is weathered to produce sedimentary rocks and, eventually, is taken back into the mantle. The hydrological cycle operates over decades and centuries as water is evaporated, taken into or released from glaciers, and flows in the great oceans currents. The third wheel represents the atmospheric cycle, operating on a scale of weeks and months, which produces the daily change in the weather. These cycles operate simultaneously, and each affects the operations of the others. Through all of this, the energy of incoming sunlight provides energy for living systems and is distributed poleward from the equator by both the oceans and the atmosphere.

With this background, the student is ready to look at environmental problems—some, like acid rain and the ozone hole, that seem well on their way to being solved, and others, such a global warming, that are yet to be dealt with in a serious manner.

Life is based on chemistry.

The behavior of molecules in living systems depends on their shape.

The first statement represents one of the greatest achievements of nineteenth-century science. We now understand that the basic reason that humans are different from other organisms is that our cells are running different chemical reactions from theirs. The stress in modern biology is on studying life at the molecular level, a fact that seems to make the subject difficult for many students. I have found it useful, therefore, to do a standard Linnaean description of life on Earth before plunging into the molecular mechanisms that make it work.

Once we are at the molecular level, the key idea is that it is atoms, not molecules, that form chemical bonds. This fact leads naturally to the notion that molecules have to be shaped so that the relevant atoms from different interacting molecules can get near each other. This isn't a hard concept for students to get, particularly if you tell them to think of the active sites as being analogous to Velcro. Once this point is made, the idea that enzymes can facilitate interactions can be made, and this, in turn, leads to the idea that it is proteins that serve as the chemical workhorses of living systems.

Life's chemistry is coded in DNA.

Students have all heard of DNA, and walking them through the familiar double helix structure introduces few problems. Both DNA replication and transcription make heavy use of the notion that it is the shape of the molecules (bases, in this case) that determines how these processes work.

I have found it best to start with a simple idea that one *gene*—one stretch of DNA—carries the code to produce one protein, which, in turn, runs one chemical reaction in the cell. Once this idea has been absorbed, it is relatively easy to add the bells and whistles that make complex cells like our own different from those of simple bacteria. This is also the place to talk about the fact that there are mechanisms in the cell that turn genes on and off, and that the process by which an organism develops involves cells turning off genes on the way to specialization. Every cell in your body, for example, has the gene for insulin, but it is turned off in all cells except for a few in the pancreas. This concept, in turn, provides a good introduction to the hot-button issues surrounding stem cells and cloning.

All living things share the same genetic code.

Besides being one of the strongest arguments for evolution, this simple fact provides the scientific basis for genetic engineering. To my mind, this statement also expresses an important truth about living systems on our planet, carrying as it does information about how they got to be the way they are and where they are likely to go in the future.

What has amazed me about the environmental debates I have heard is how little attention is being paid to the idea that, because of human action, the genetic makeup of plants and animals in the near future may be quite different from what it is now. Biologists talking about the response of plants to higher atmospheric carbon dioxide, for example, have shown that some plants slow down their carbon uptake when a lot of the material is available—in essence, they get "lazy." That may be true of plants that evolved naturally, but it's easy to imagine engineering trees in which the "laziness" gene or genes don't operate. Such a development would be a reversal of traditional agricultural practices, in which plants are bred to be efficient, using as little carbon as possible, but such carbon-hungry plants could easily be grown as a way of pulling carbon dioxide out of the air. This possibility is just one example of the kind of impact genetic engineering might have on our future.

Look at it this way: Ten thousand years ago, our ancestors decided that they would not be content with what nature offered freely in the way of food, and began the practice of agriculture to free themselves from hunting-gathering. In a similar way, we are now approaching the point where we will not have to be content with what nature offers us in the way of genetic material, but can modify it to suit our needs.

This way of thinking raises enormous ethical, moral, religious, and legal issues, of course, which makes it an ideal subject to bring up toward the end of a scientific literacy class.

Life evolved through the process of natural selection.

This is one of those ideas that run through an entire branch of science, unifying seemingly diverse concepts. It is important to understand that there are actually two steps in the evolutionary process, and the classical Darwinian theory deals only with one of them.

The two steps are (1) chemical evolution (in which inorganic materials in the early Earth gave rise to the first living cell), and (2) evolution by natural selection (in which that cell and its descendants produced the millions of species we see around us today). The latter of these areas is the subject matter of a mature field of science, with theories supported by massive amounts of fossil and other data. The former is a newer field, having received serious research funding only in the last decade or so. It is also, to my mind, one of the hottest fields of current research anywhere in the sciences.

Many controversies arise from evolutionary theory. There is, for example, the perceived conflict with religious doctrines, which gives rise to legal battles over the teaching of creationism or Intelligent Design in the schools. Many individuals, mainly fundamentalist Protestants, have deep reservations about the impersonal nature of the process and its seeming denial of divine purpose. Most mainline Christian and Jewish theologians, however, including the late Pope John Paul II, have found it possible to reconcile their faith with the fact that life has evolved. (In fact, I have found the statement "If it's good enough for the Pope, it's good enough for me" to be an effective way of dealing with objections raised on this score during question periods after my talks.)

On a more secular level, scientists are starting to interpret modern human behavior in the light of our evolutionary past. This has given rise to fields with names like evolutionary psychology and even (God help us!) evolutionary marriage counseling. I find some of the more popular versions of this trend to be something of an exercise in constructing "just so" stories, but at its core it is, I think, leading us to a new vision of what it means to be a human being.

THE BEST IS THE ENEMY OF THE GOOD

I would argue that people who have acquired knowledge of these Great Ideas have, in fact, been supplied with precisely the kind of educational foundation I discussed at the start of this chapter and will therefore be, by my definition, scientifically literate. Those people will have the intellectual framework that will allow them to understand the basic science behind whatever issues arise. This does not mean, however, that they will have the ability to think about these issues in the same critical, quantitative way that a scientist would, and this, of course, raises an important question: Is the sort of foundation I've described really enough?

This question almost always arises when reform of the science curriculum is on the table, usually in the context of the methods versus content debate outlined in Chapter 10. As discussed there, there seems to be a strong prejudice, especially among academic scientists, that only the most advanced kind of science really matters, and that presenting anything else amounts to giving the students a watered-down education. I characterized this, tongue in cheek, as the "everybody should have a Ph.D. in everything" school of thought.

Now that I have laid out what I think is the proper foundation for science education, I would like to return to this issue for a moment.

The goal of trying to turn every citizen into a miniature scientist, it seems to me, is not only completely impractical, but also pointless. As I said in Chapter 3, one of the "dirty little secrets" of the scientific community is that research-level scientists are themselves often completely illiterate in scientific areas outside of their own specialty. Thus training someone to think like a physicist isn't going to do a lot of good if that person has to deal with an issue involving molecular biology. As far as civic participation is concerned, breadth trumps depth any time.

In addition, as I pointed out in the last chapter, the nature of science itself is changing, and issues in the future are likely to involve the kinds of massive computational exercises we outlined for global warming. Only a few specialists are going to have the ability to root around inside those computer codes and determine things like their accuracy and validity. This means that the rest of us—and that includes most scientists—will have a very different role from that envisioned by those who expect people to form their own opinions on scientific issues. The idea that they should be able to do so, I think, is simply a holdover of John Dewey's notion of a scientific habit of mind, discussed in Chapter 8, and is completely inappropriate for the twenty-first century.

THE ROLE OF LABORATORIES AND DISCOVERY

One aspect of the attempt to turn students into scientists is the notion that students should, in some sense, mimic the scientific discovery process in their learning experience. Going by names like Discovery or Inquiry-based Learning, this philosophy relies heavily on laboratory and fieldwork to introduce students to science.

Before I go on, let me make a couple of comments. First, I am aware that education, like every other field, is subject to fads; as a grad student, I calculated the half-life of fads in my own field of particle physics to be roughly 2 years. It is in the nature of fads that they generate intense but transitory levels of enthusiasm and that this enthusiasm is rarely based on sound data. I believe that I can see certain elements of faddishness in the current excitement about inquiry-based learning.

Second, I am perfectly comfortable with the notion that this type of learning plays well into the interests and attention spans of young students. I suspect, for example, that it is a very good way to get elementary and middle school students interested in science. At the university level, however, which is where I spend most of my time, it tends to play out in a kind of knee-jerk statement such as "every student must take two semesters of laboratory science" whenever area requirements in science are discussed.

If the goal of university education is to turn out scientifically literate graduates, I would argue that laboratory experience is really something of an unnecessary frill, a throwback to an earlier, simpler age. It seems to me that scientific literacy can most easily be taught in the standard lecture-classroom format, without recourse to laboratories at all. I would even go so far as to argue that at the university level laboratory classes for nonscientists don't do much good at all, and may actually do harm. I would love to be proved wrong about this, but when I push people to justify the traditional laboratory requirement in science, the most common response tends to be some variant of "we've always done it that way." (Another common response, somewhat discouraging but probably more honest, is that all those lab sections generate much-needed support for graduate students.)

I suppose my skepticism on the subject of labs began when I was a young associate professor at the University of Virginia. I taught graduate courses at the time, so I was largely invisible to undergraduate students. I would occasionally wander down into the lab rooms where the students, if they noticed me at all, probably assumed I was from Building and Grounds (like many physicists, I never wear a coat and tie to work). What I overheard on those forays was deeply disturbing, because I found that fully a third of the students were working their way backward from the "right" answers to the entries to put into their data logs. What these students were learning, then, was not the scientific method, but almost an antiscience. Their response was, I suppose, a rational response to a situation in which they knew they would be graded on whether their results matched the "right" answers. Nevertheless, I suspect that they are now adults who are highly skeptical of anything they hear from the scientific community. This is not a good outcome.

I am aware that it is possible that not all university labs operate this way (although I challenge senior faculty, at least those who are not easily shocked, to do what I did at their own institutions). Nevertheless, I believe that there are factors in the structure of universities that militate against laboratories—even well-designed laboratories—contributing to scientific literacy in any meaningful way.

The primary reason for this opinion involves the sheer logistical problem of moving hundreds of students through lab sections every week. There are basically two ways to do this: (1) You can have simple, cheap experiments; or (2) you can do computer simulations of experiments. Neither option, I think, is very useful.

The requirement that lab equipment be kept working through the course cycle places severe restrictions on the kinds of experiments that can be done. Typically, the more complex and realistic the experiment is, the more likely it is that the equipment will break down and induce frustration in both students and instructors. The only way to get around this without incurring high personnel costs is to use experiments that require simple, idiot-proof equipment. This, in turn, means that the experiments will have to be of the "roll the ball down the inclined plane" variety. As I argued in the previous chapter, this sort of learning experience will do nothing to equip students to deal with the kinds of complex public issues that science will be raising in the future.

The other option—to have the experiments done by computer simulation—is equally unsatisfactory, at least in the versions I've seen and worked with. Instead of dissecting a frog, for example, students look at a computer screen. Instead of timing the descent of a dropped ball, they watch a ball on a screen and read the numbers from a display.

Such simulations have the advantage of not requiring complex equipment, but think for a moment about what the student is learning. They do not learn that the real world is out there, to be measured and observed. Instead, it appears magically on a screen, like a (rather dull) video game. The bottom line: You have to believe what the computer tells you. I can't imagine a worse kind of background knowledge to impart to students who will have to deal with issues like judging the validity of computer models.

My conclusions, then, are that simple labs, including computer simulations, don't prepare students for the world they will encounter when they leave school, and more realistic, consequently more complex, labs are unlikely to work in the context of large introductory science courses. In the end, I would argue that the time and money now spent on make-work laboratories would be much better spent on finding other ways to increase student scientific literacy—by training faculty to teach integrated science literacy courses, for example.

THE NEXT STEP

According to Jon Miller's data, which I discussed in Chapter 6, we are making progress in giving students the basic foundation in scientific literacy that they will need to function as citizens. Today, some 28% of Americans pass his (admittedly basic) standards of scientific literacy.

This is an encouraging and challenging statistic. It is encouraging because more and more Americans are acquiring a minimal foundational knowledge of science. It is challenging because it means that there is now a sizable population that is ready to take the next step and go beyond simple scientific literacy—in effect, to start building the superstructure of the house on the foundation that is already in place. And this, in turn, leads us to an intriguing question: What would that second step look like?

Before I turn to this question, it has to be emphasized that the fact that roughly a quarter of the American population can be called scientifically literate means that three quarters cannot. We can never forget that the main emphasis on educational reform has to remain on basic scientific literacy. This is an important point, because there is always a strong temptation for educators to concentrate attention on their best students—in this case, those ready for the second step—and ignore that great majority left behind. This temptation must be resisted.

We also have to keep a tight rein on our enthusiasm as we approach this issue. The fact that someone knows the difference between an atom and a molecule, for example, doesn't mean that he or she is ready to start reading *Science* or *Nature* (or, for that matter, *Scientific American*). Meeting Jon Miller's criteria for scientific literacy, in other words, does not automatically turn someone into a candidate for a career in science, or even into someone who has acquired a scientific habit of mind. We have to recognize the fact that we cannot turn everyone into a scientist, and that any attempt to do so can only lead to certain failure.

So what is a realistic goal for the second step? Perhaps we can get a hint of what this might look like by considering two groups who routinely deal with complex issues in science: scientists themselves, and members of the legal profession. As I pointed out in Chapter 3, there is nothing in the training of scientists that prepares them to deal with issues outside of their field of specialization. How, then, do they form opinions on issues outside of their areas of expertise?

When I discussed this issue with Jon Miller, he had an interesting comment. "Look," he said, "when you and I want to know about some issue, we go to our Rolodex and call somebody in the field. The problem is that the average person doesn't have our Rolodex." I call this the Golden Rolodex theory of information acquisition.

After this conversation with Miller, I began conducting an informal survey among my colleagues as I traveled around the country. I would ask them how they came to decisions about issues outside of their own fields. As Miller had predicted, the answers almost always involved the Golden Rolodex. Some samples: "I call a friend in the field," "I go with the consensus," "I call someone at the home institution (of the proponent of a controversial idea) to get a read on him," and so on. In cases of conflict between experts, the Golden Rolodex can be used to render a judgment about which of the competing experts is more likely to be right, and which ought to be ignored.

I can see the Golden Rolodex operating in my own life. When I have a question about a complex problem (global warming is a good example), I don't try to sit down and write a computer code to simulate the Earth's climatic future; I have neither the time nor the expertise to do something like that. Instead, I call on friends whose opinions span the entire spectrum of debate, people located at universities around the country. This personal panel of experts gives me a sense of what is known about the question I'm asking, what the points of contention are, and how they are likely to be resolved. My informal survey seems to indicate that most other scientists form their opinions in much the same way, by discussion with colleagues more knowledgeable in the particular field at issue. In the end, I think, most of us come to a decision based on which of our friends we trust the most.

This isn't a bad way to proceed. In fact, the law operates in much the same way, albeit acting according to a much more formal set of rules. It often happens that legal disputes hinge, at least in part, on scientific questions: Did this chemical cause the plaintiff's disease? Was this product defective? In these cases there is a clearly laid out procedure for bringing expert knowledge before the court. The attorneys in the case bring in witnesses who, by education or training, can be qualified as *experts*. Once this has been done, there are well-established precedents by which the judge examines the testimony they are to give and decides whether it can be presented to a jury. This judgment can often be the result of a complex legal proceeding, almost a minitrial in some cases. Only after this screening has been done is the testimony of the expert presented to a jury.

The criteria by which these judgments are made were first laid out for the federal courts in 1993 in the case *Daubert v. Merrell Dow Pharmaceuticals*. Justice Blackmun, writing for the Supreme Court, said that henceforward it would be the responsibility of the judge to ensure that testimony presented to the court was "not only relevant, but reliable." [1] He then laid out a series of criteria for making this kind of judgment. Was the work presented in the testimony peer reviewed? Was it generally accepted in the scientific community? Could the claims be tested? These kinds of criteria (in addition to other, more technical considerations) are actually a pretty close match to the kinds of things that scientists consider when evaluating a new theory or idea.

It is important to realize that it is quite common for the testimony of experts from both sides in a trial to pass these tests and be presented to a jury. In this case, the jury will very likely hear qualified experts testifying to diametrically opposite "truths." Thus the jury has to decide which expert to believe, and this process often involves nonscientific criteria, such as which witness seemed more believable. The jury, in other words, is in the same position as the average citizen reading about conflicting scientific claims in the newspaper. Seemingly equally qualified experts contradict each other, and it's up to the nonexpert reader to decide which to believe.

There are some general conclusions we can draw by looking at the way that the scientific and legal communities approach issues of scientific validity:

1. Experts in one field usually need to rely on experts in other fields.
2. All experts are not created equal, and in cases of conflict between experts, at least one is likely to be wrong.
3. Deciding which expert to believe may not be a completely rational process.

It seems to me that the proper aim of the second step in general science education should not be to turn all citizens into miniature scientists capable of analyzing data and coming to their own opinions on important issues; indeed, such a goal is beyond the reach of scientists themselves. Instead, the goal should be to (1) provide the training needed to screen out nonreliable claims, and (2) give the person enough background to make a decision as to which expert to trust.

Having said all this, we have to recognize that people being what they are, this process will never meet rigid tests of rationality. Inevitably, there will be people who believe an expert because he's wearing a nice tie or because she has a great smile. Human beings are not now, and never will be, totally rational, a fact that has long been recognized in the legal system. The best we can do is to give citizens the tools that they need to make judgments on complex issues, and then get out of the way and let them choose.

Some of my colleagues are very uncomfortable with this notion. "If only these people knew what I know," they say, "they would come to the same conclusions that I do." I'm afraid that it just doesn't work that way. My experience, for example, is that most people who oppose the use of embryonic stem cells know perfectly well what a stem cell is—they just come to a different conclusion on the subject than most scientists (including the author). In the end, if you want to live in a democracy, you have to accept the fact that you will sometimes be outvoted, not matter how much education you have.

A FINAL WORD

Throughout this essay I have tried to emphasize the importance of producing a scientifically literate citizenry. Whether you think my own approach, outlined above, is the right one to follow or not, there can be no questioning that it is time for us to get to work on this issue. In the end, I think Carl Sagan put it best in the quote with which we closed Chapter 3:

> We've arranged a global civilization in which the most crucial elements. . . profoundly depend on science and technology. We've also arranged things so that no one understands science and technology. This is a prescription for disaster.

Notes

Chapter 1

1. Quoted in J. Trefil, *Reading the Mind of God* (New York: Charles Scribners & Sons, 1989), p. 18.

2. K. Boulding, *The Image: Knowledge in Life and Society* (Ann Arbor: University of Michigan Press, 1956).

3. John E. Jones's ruling in *Kitzmiller v. Dover Area School Board* (December 20, 2005) is available online at http://www.pamd.uscourts.gov/kitzmiller/kitzmiller_342.pdf.

Chapter 3

1. C. Sagan, *The Demon-Haunted World: Science as a Candle in the Dark* (New York: Random House, 1995).

Chapter 4

1. C. P. Snow, *Two Cultures and the Scientific Revolution* (Cambridge: Cambridge University Press, 1959).

2. F. R. Leavis, quoted in D. P. Barash, "C. P. Snow: Bridging the Two-Cultures Divide," *The Chronicle of Higher Education, 52* (2005), B10–11.

3. S. Johnson, quoted in J. Boswell, *Life of Samuel Johnson* (Book 3), available online at http://www.samueljohnson.com/refutati.html#57.

Chapter 5

1. T. Burnet, *The Sacred Theory of the Earth, in Which Are Set Forth the Wisdom of God Displayed in the Works of the Creation, Salvation, and Consummation of All Things . . .* (London: T. Kinnersley, 1816).

2. A. Eddington, quoted in A. V. Douglass, *The Life of Arthur Stanley Eddington* (London: Nelson, 1956), as cited online at http://www.groups.dcs.stand.ac.uk/~history/Biographies/Eddington.html.

3. W. Wordsworth, "The Tables Turned: An Evening Scene On The Same

Subject," *The Complete Poetical Works* (London: Macmillan, 1888; Bartleby. com, 1999). Available online at http://www.bartleby.com/145/ww134.html

4. W. Whitman, "When I Heard the Learn'd Astronomer," in *Leaves of Grass* (Philadelphia: David McKay, 1900; Bartleby.com, 1999). Available online at http://www.bartleby.com/142/180.html.

5. W. Blake, "Milton," in D. H. S. Nicholson and A. H. E. Lee, (Eds.), *The Oxford Book of English Mystical Verse* (Oxford: Clarendon Press, 1917; Bartleby. com, 2000). Available online at http://www.bartleby.com/236/62.html

6. W. Gilbert and A. Sullivan, "If You're Anxious to Shine," from *Patience* (New York: G. Shirmer, n.d.; original published in 1881).

7. For an explanation and photograph of this painting, see Newdoll's Web site, Brush with Science (http://www.brushwithscience.com/Spring2003/LifeForms2003.html).

Chapter 6

1. J. Miller, "The measurement of civic scientific literacy," in *Public Understanding of Science, 7,* 203–223; and "Civic scientific literacy across the life cycle," paper presented at the annual meeting of the American Association for the Advancement of Science, San Francisco, 2007.

2. J. Miller, "Public understanding of science: Are Europeans better at it?" Paper presented at the annual meeting of the American Association for the Advancement of Science, San Francisco, 2007.

Chapter 7

1. B. Franklin [R. Saunders, pseudo.], "How to secure Houses, &c. from Lightning," in *Poor Richard Improved: Being an Almanack and Ephemeris... for the Year of our Lord 1753* (Philadelphia: B. Franklin and D. Hall, 1753). Available online at http://www.franklinpapers.org/franklin/framedVolumes. jsp?vol=4&page=403a.

Chapter 8

1. H. Spencer, "What Knowledge Is of Most Worth," in *Essays on Education, etc.* (London: J. M. Dent and Sons, 1911).

2. T. Huxley, "Science and Culture," in *Science and Education* (New York: P. F. Collier and Son, 1964/1880).

3. A. de Tocqueville, *Democracy in America* (New York: Vintage Books, 1961/1835).

4. D. C. Gilman, *A Brief History of Jhu* [Johns Hopkins University] ([1876]). Available online at http://webapps.jhu.edu/jhuniverse/information_about_hopkins/about_jhu/a_brief_history_of_jhu/.

5. Ibid.

6. J. Dewey, "Symposium on the purpose and organization of physics teaching in secondary schools (part 13)," in *School Science and Mathematics, 9* (1909), 291–292.

7. Ibid.

8. I. C. Davis, "The measurement of scientific attitudes," in *Science Education, 19* (1935), 117–122.

9. L. Hill, *Hearings before the Committee on Labor and Public Welfare, United States Senate: Science and Education for National Defense* (Washington, DC: U.S. Government Printing Office, 1958).

10. M. H. Shamos, *The Myth of Scientific Literacy* (New Brunswick, NJ: Rutgers University Press, 1995).

11. J. Zacharias, quoted in J.L. Rudolph, "PSSC in Historical Context: Science, National Security, and American Culture During the Cold War." Available online at http://www.compadre.org/portal/pssc/docs/Rudolph. pdf.

12. G. Seaborg, quoted in Koret Task Force on K–12 Education, *Our Schools and Our Future . . . Are We Still at Risk?* (Stanford, CA: Hoover Institution Press, 2003). Available online at http://media.hoover.org/documents/0817939210_3.pdf.

Chapter 9

1. S. Rimer, "Harvard Task Force Calls for New Focus on Teaching and Not Just Research," *New York Times,* May 10, 2007.

Chapter 10

1. P. Medawar, *Pluto's Republic* (New York: Oxford University Press, 1982).

2 . R. March, *Physics for Poets* (New York: McGraw-Hill, 1996).

Chapter 11

1. J. Searle, personal communication.

Chapter 12

1. H. Blackmun, in *Daubert v. Merrell Dow Pharmaceuticals, Inc.*, 509 U.S. 579 (1993), Section II B. Available online at http://caselaw.lp.findlaw.com/scripts/getcase.pl?court=us&vol=509&invol=579.

References

Alvarez, L. W. (1987). *Alvarez: Adventures of a physicist*. New York: Basic Books.

Asia Society. (2006, May). *Math and science education in a global age: What the U.S. can learn from China*. New York: Asia Society, Education Division.

Barash, D. P. (2005, November 25). C.P. Snow: Bridging the two-cultures divide. *The Chronicle of Higher Education, 52*(14), B10–11.

Blake, W. (1917). Milton. In D. H. S. Nicholson & A. H. E. Lee (Eds.), *The Oxford book of English mystical verse*. Oxford: Clarendon Press. Available online at http://www.bartleby.com/236/62.html.

Boulding, K. (1956). *The image: Knowledge in life and society*. Ann Arbor: University of Michigan Press.

Boyd, B. (2006, Autumn). Getting it all wrong: Bioculture critiques cultural critique. *The American Scholar, 75*(4), 18–32.

Burnet, T. (1816). *The sacred theory of the earth, in which are set forth the wisdom of God displayed in the works of the creation, salvation, and consummation of all things. . . .* London: T. Kinnersley. (Original work published in Latin, with title: *Telluris theoria sacra*, 1681–89)

Darwin, C. (1859). *On the origin of the species by means of natural selection: or, The preservation of favoured races in the struggle for life*. London: J. Murray.

Daubert v. Merrell Dow Pharmaceuticals, Inc., 509 U.S. 579 (1993), Section II B. Available online at http://caselaw.lp.findlaw.com/scripts/getcase.pl?court=us&vol=509&invol=579.

Davis, I. C. (1935). The measurement of scientific attitudes. *Science Education 19*, 117–122.

Dewey, J. (1909). Symposium on the purpose and organization of physics teaching in secondary schools (part 13). *School Science and Mathematics 9*, 291–292.

Dewey, J. (1933). *How we think: A restatement of the relation of reflective thinking to the educative process*. Boston: Heath.

Douglass, A. V. (1956). *The life of Arthur Stanley Eddington*. Londnon: Nelson. Cited online at http://www.groups.dcs.stand.ac.uk/~history/Biographies/Eddington.html

Franklin, B. [R. Saunders, pseudo.]. (1753). How to secure houses, &c. from lightning. *In Poor Richard improved: Being an almanack and ephemeris...for the year of our Lord 1753*. Philadelphia: B. Franklin and D. Hall. Available online at http://www.franklinpapers.org/franklin/framedVolumes.jsp?vol=4&page=403a.

Gereffi, G., & Wadhwa, V. (2005, December). *Framing the engineering outsourcing debate: Placing the United States on a level playing field with China and India.* Durham, NC: Duke University, Master of Engineering Management Program. Available online at http://memp.pratt.duke.edu/downloads/duke_outsourcing_2005.pdf.

Gilbert, W., & Sullivan, A. (1881). If you're anxious to shine [song]. In *Patience.* New York: G. Shirmer.

Gilman, D. (1876). *A brief history of JHU [Johns Hopkins University].* Retrieved May 3, 2007, from http://webapps.jhu.edu/jhuniverse/information_about_hopkins/about_jhu/a_brief_history_of_jhu/

Gosse, P. H. (1857). *Omphalos: An attempt to untie the geological knot.* London: J. Van Voorst.

Hill, L. (1958). *Hearings before the Committee on Labor and Public Welfare, United States Senate: Science and education for national defense.* Washington, DC: U.S. Government Printing Office.

Hirsch, E. D. (1987). *Cultural literacy: What every American needs to know.* Boston: Houghton-Mifflin.

Hirsch, E. D., Kett, J., & Trefil, J. S. (2002). *The new dictionary of cultural literacy.* Boston: Houghton-Mifflin. (Originally published in 1988 as *The dictionary of cultural literacy*).

Huxley, T. H. (1964). Science and culture. In *Science and education.* New York: P. F. Collier and Son. (Original work published 1880)

Kitzmiller v. Dover Area School Board (2005). Available online at http://www.pamd.uscourts.gov/kitzmiller/kitzmiller_342.pdf.

Koret Task Force on K–12 Education. (2003). *Our schools and our future . . . Are we still at risk?* Stanford, CA: Hoover Institution Press. Available online at http://media.hoover.org/documents/0817939210_3.pdf

Kusche, L. (1995). *The Bermuda Triangle mystery solved.* Amherst, NY: Prometheus Books. (Original work published by Harper & Row in 1975)

Lomborg, B. (2001). *The skeptical environmentalist: Measuring the real state of the world.* New York: Cambridge University Press.

March, R. (1996). *Physics for poets* (4th Ed.). New York: McGraw-Hill.

Medawar, P. (1982). *Pluto's republic.* New York: Oxford University Press.

Miller, J. (1997). The measurement of civic scientific literacy. *Public Understanding of Science, 7,* 203–223.

Miller, J. (2007a, February). *Civic scientific literacy across the life cycle.* Paper presented at the annual meeting of the American Association for the Advancement of Science, San Francisco.

Miller, J. (2007b, February). *Public understanding of science: Are Europeans Better at it?* Paper presented at the annual meeting of the American Association for the Advancement of Science, San Francisco.

Miller, J., Pardo, R., & Niwa, F. (1997). *Public perceptions of science and technology: A comparative study of the European Union, the United States, Japan, and Canada.* Bilbao: Fundacion BBV.

Morowitz, H. J., & Trefil, J. S. (1992). *The facts of life: Science and the abortion controversy.* New York: Oxford University Press.

National Commission on Excellence in Education. (1983). *A nation at risk: The imperative for educational reform.* Washington, DC: U.S. Department of Education.

No Child Left Behind Act of 2001. Pub. L. No. 107-110, 115 *Stat.* 1425 (2002).

Preston, R. (2000, January 12). The genome warrior. *The New Yorker*, pp. 66–83.

Raup, D. M., & Sepkoski, J. J. (1984). Periodicity of extinctions in the geologic past. *Proceedings of the National Academy of Sciences of the United States of America, 81*(3), 801–805.

Sagan, C. (1995). *The demon-haunted world: Science as a candle in the dark.* New York: Random House.

Shamos, M. H. (1995). *The myth of scientific literacy.* New Brunswick, NJ: Rutgers University Press.

Sokal, A. (1996, May). A physicist experiments with cultural studies. *Lingua Franca, 6,* 62–64. Available online at http://www.physics.nyu.edu/faculty/sokal/lingua_franca_v4/lingua_franca_v4.html.

Sokal, A. (1996, Spring/Summer). Transgressing the Boundaries: Towards a Transformative Hermeneutics of Quantum Gravity. *Social Text*, No. 46/47, 217–252. Available online at: http://www.physics.nyu.edu/faculty/sokal/transgress_v2/transgress_v2_singlefile.html.

Spencer, H. (1911). *Essays on education, etc.* London: J. M. Dent and Sons.

Snow, C. P. (1959). *Two cultures and the scientific revolution.* Cambridge, UK: Cambridge University Press.

Tocqueville, A. (1961). *Democracy in America.* New York: Vintage Books. (Original work published 1835)

Trefil, J. S. (1997). *Are we unique? A scientist explores the unparalleled intelligence of the human mind.* New York: Wiley & Sons.

Trefil, J. S. (2004). *Human nature: A blueprint for managing the earth—by people, for people.* New York: Henry Holt.

Trefil, J. S., & Hazen, R. M. (2007). *The sciences: An integrated approach* (5th Ed.). New York: Wiley. (Original work published 1995)

Von Baeyer, H. C. (1998). *Maxwell's demon: Why warmth disperses and time passes.* New York: Random House.

Weinberg, S. (1992). *Dreams of a final theory.* New York: Pantheon Books.

Whitman, W. (1900). When I heard the learn'd astronomer. In *Leaves of grass.* Philadelphia: David McKay. Available online at http://www.bartleby.com/142/180.html.

Wordsworth, W. (1888). The tables turned: An evening scene on the same subject. *The complete poetical works.* London: Macmillan. Available online at http://www.bartleby.com/145/ww134.html.

Index

About the Author

James Trefil is Clarence J. Robinson Professor of Physics at George Mason University. He received his education at the University of Illinois and as a Marshall Scholar at Oxford before finishing his graduate work in physics at Stanford. His research career in physics resulted in more than 100 papers in professional journals.

He has been a prominent contributor to public education in science, and has been the author or coauthor of almost 40 books. He has also served as science advisor and contributor to many institutions, including *Smithsonian Magazine,* National Public Radio, and the Adler Planetarium in Chicago. His interest in scientific literacy began with his participation as coauthor, with E. D. Hirsch and Joseph F. Kett, of *The Dictionary of Cultural Literacy.*

His current interests include lecturing and writing about the intersection of science and the law and incorporating the Great Ideas approach to scientific literacy in curricula from middle school to university levels. He lives with his wife in Fairfax, Virginia.